ENVIRONMENTAL RESOURCES
and
APPLIED WELFARE ECONOMICS

CONTRIBUTORS

Gardner M. Brown, Jr., Department of Economics,
University of Washington

Charles J. Cicchetti, Energy and Environmental Policy Center,
John F. Kennedy School of Government, Harvard University

Steven E. Daniels, Department of Forest Resources,
Utah State University

Robert K. Davis, Institute of Behavioral Science,
University of Colorado at Boulder

Anthony C. Fisher, Department of Agricultural and
Resource Economics, University of California, Berkeley

A. Myrick Freeman III, Department of Economics,
Bowdoin College

Robert H. Haveman, Department of Economics,
University of Wisconsin, Madison

Charles W. Howe, Institute of Behavioral Science,
University of Colorado at Boulder

William F. Hyde, School of Forestry and
Environmental Studies, Duke University

Allen V. Kneese, Resources for the Future

Jeffrey C. Norris, Dover-Foxcroft, Maine

Talbot Page, Center for Environmental Studies and
Center for Public Policy, Brown University

V. Kerry Smith, Department of Economics and Business,
North Carolina State University

Joseph Swierzbinski, Department of Economics and
School of Natural Resources, University of Michigan, Ann Arbor

Elizabeth A. Wilman, Department of Economics,
University of Calgary

ENVIRONMENTAL RESOURCES
and
APPLIED WELFARE ECONOMICS

Essays in Honor of

John V. Krutilla

V. KERRY SMITH
Editor

RESOURCES FOR THE FUTURE • Washington, D.C.

Printed in the United States of America

Published by Resources for the Future
1616 P Street, N.W., Washington, D.C. 20036

Books from Resources for the Future are distributed worldwide by The Johns Hopkins University Press.

Library of Congress Cataloging-in-Publication Data
Environmental resources and applied welfare economics: essays in
    honor of John V. Krutilla / V. Kerry Smith, editor.
        p.   cm.
    "Selected bibliography of works by John V. Krutilla": p.
    Includes index.
    ISBN 0-915707-40-3
    1. Environmental policy.   2. Natural resources.   3. Welfare
economics.   4. Krutilla, John V.   I. Krutilla, John V.   II. Smith,
V. Kerry (Vincent Kerry), 1945–   .
HC79.E5E585 1988
333.7—dc19   88-21808
                                                                       CIP

This book was edited by Dorothy Sawicki and designed by Debra Naylor. The index was prepared by Florence Robinson.

∞ The paper in this book meets the guidelines for permanence and durability of the Committee on Production Guidelines for Book Longevity of the Council on Library Resources.

CONTENTS

Foreword   ix
Preface and Acknowledgments   xiii

PART I   Background

The Influence of Resource and Environmental
Problems on Applied Welfare Economics:
An Introductory Essay   3
V. Kerry Smith

Three Decades of Water Resources Research:
A Personal Perspective   45
Allen V. Kneese

PART 2   Resource Problems and Theoretical Issues in Applied Welfare Economics

Key Aspects of Species Extinction:
Habitat Loss and Overexploitation   59
Anthony C. Fisher

Intergenerational Equity and the Social
Rate of Discount   71
Talbot Page

Optimal Genetic Resources in the Context of
Asymmetric Public Goods   91
Gardner M. Brown, Jr., and *Joseph Swierzbinski*

**Quasi-Optimal Pricing for Cost Recovery in
Multiple Purpose Water Resource Projects**   119
A. Myrick Freeman III and *Jeffrey C. Norris*

**Balancing Market and Nonmarket Outputs on
Public Forest Lands**   135
William F. Hyde and *Steven E. Daniels*

PART 3   Resource Policies and the Practice of Applied Welfare Economics

**Modeling Recreation Demands for Public
Land Management**   165
Elizabeth A. Wilman

**Public Intervention Revisited:
Is Venerability Vulnerable?**   191
Charles W. Howe

**Lessons in Politics and Economics
from the Snail Darter**   211
Robert K. Davis

**Environmental Litigation and Economic
Efficiency: Two Case Studies**   237
Charles J. Cicchetti and *Robert H. Haveman*

**Appendix: "Conservation Reconsidered,"
by John V. Krutilla**   261

**The Works of John V. Krutilla:
A Bibliographic Profile**   275

Index   281

FOREWORD

This volume, a festschrift for John V. Krutilla, celebrates the contributions to the study of applied welfare economics by one of the nation's leading economists, a long-time member of the research staff at Resources for the Future (RFF). The authors of the papers are John's friends and colleagues. They have ample reason to mark his career, for he has been an intellectual and a personal mentor to them.

John has been at RFF for thirty-three of its thirty-six years, long enough for each to have acquired many of the characteristics of the other. He has devoted his career here to a few large problems—water resources, the economics of natural environments, and, most recently, multiple use land management. In each case John has produced extraordinary research, and more. Invariably, his vision of the problem has been ample enough to encourage other scholars to share productively in it. And what he and they have done has made a difference in the world.

For this latter reason, if for no other, it is appropriate that these writings have importance beyond their celebratory purpose. Indeed, they help us understand how economic scholarship has shaped the standards that we as a society use in deciding how to develop and conserve our natural and environmental resources. It is in forging these standards that John Krutilla carried out some of his most fundamental work.

Roughly speaking, applied welfare economics concerns itself with wise use of the resources in which society at large has a significant stake. Scholars in this field argue that this stake involves *public goods that have value,* as does private property, and that *efficiency is an important goal* in the management of public resources. These principles seem harmless enough until one begins to use them, for their application influences policy in ways that are sometimes less than popular.

Value is created when public resources are put to public or private use. Public uses such as water resource projects and recreation on public lands were the first to be analyzed by economists. Later, private activity and its impact on societal values came under study, particularly with research on environmental pollution. More recently, new kinds of values are being added to the research agenda. They are represented in this volume by discussions of species extinction and the preservation of genetic resources. And values ascribed to future generations are being included with our own as standards for judging policy. Thus, intergenerational equity has become a central concern in today's debates about hazardous waste disposal and sustainable development.

The reason for this preoccupation with value is that the wise use of resources requires knowing both the public and the private costs involved. Thus, "value" in this context is not an abstraction. It requires that scholars in this field put numbers to the things that society values and that they compare these public values with the private ones set by markets. It forces these scholars to be attentive to, and if possible to anticipate, the public's evolving concerns about how we use our resources. However elegant the scholarship, this is highly pragmatic work.

The other economic principle that shapes our view of resources policy is that of efficiency. Markets allocate private goods efficiently, and it seems reasonable to manage public resources with equal care. The problem, of course, is that there are no market prices for public goods. As a result, either no markets exist to allocate these resources efficiently or, absent prices for them, private markets fail to account for their true value.

In application, these two principles involving value and efficiency direct public policy about natural resources and the environment in ways bound to cause controversy. For example, they lay the conceptual basis for government intervention in markets to ensure that public values are accounted for. More-over, because efficiency is generally achieved by pricing resources accurately, these principles suggest that we should pay for the use of public resources. Such payments may range from user fees for recreation to investments in pollution control equipment. Not much imagination is needed to think of interests that variously applaud or deplore these results.

But value and efficiency are not the only standards for public policy, as advocates for them soon discover. Public policy is also shaped by goals as disparate as achieving fairness in income distribution, preserving natural habitats for their own sake, keeping faith with one's constituents, and honoring political accommodations. The resulting tension is often characterized as a conflict between efficiency and equity. This is a misleading shorthand, however, unless one defines equity as everything that a pluralistic democracy is about—except efficiency.

Considering that efficiency is only one of several standards for policy, it is little wonder that economists frequently point with alarm to examples of inefficiency in the use of public resources. Several of the chapters in the volume are firmly in this tradition. Still, it would be wrong to conclude that efficiency does not matter in public policy just because it does not always prevail. The remarkable thing is that the economist's view has developed a strong, clear voice in the debate, an accomplishment due in no small part to John Krutilla. And it seems to me inevitable that federal budget stringency and growing demands on our natural and environmental resource base can only create more attentive listeners to that voice.

This volume thus deepens our understanding of the close linkage of scholarship and policy, and that is an important contribution. But whatever its other virtues, a festschrift should first convey a sense of celebration. And so this one does. John Krutilla's scholarship over the years reassures us that ideas matter, for his certainly have. The admiration of his colleagues that this book embodies reminds us that people also matter—even, or maybe especially, in the world of ideas. These affirmations of John's scholarship and humanity are what these celebratory writings are really about.

August 1988   ROBERT W. FRI
                                                              President
                                               Resources for the Future

PREFACE
AND
ACKNOWLEDGMENTS

This volume was conceived in an effort to honor John V. Krutilla on the occasion of his sixty-fifth birthday. Those of us who prepared papers for the volume, and many others as well, have been captured by his ideas. We have been influenced by John's own work over the course of his more than thirty years at Resources for the Future, as well as by the opportunity we each had to work with him, directly or indirectly.

Good research is at least partly a science of persuasion—a matter of convincing our peers that we have something important, or at least useful, to say about a problem. John has done more than that. He has opened our eyes to new problems. He has shown us that economic analysis has something important to say about natural environments, wildlife, wildlands, and scenic resources. Moreover, the features of these problems as John identified them offer equally important insights as we look anew at the familiar problems confronting public investment decisions. Irreversibilities, uncertainty, jointness in both production and consumption, in addition to the complexities of defining consumers' valuation, accompany many of the important policy issues of today outside the area of resource economics.

On the eve of publication of his latest book (with Michael D. Bowes) on the economics of multiple use land management, it would be premature to summarize or to attempt to take stock of John's work. Even though he is retiring from Resources for the Future, we know that his research is certainly not finished. Rather, we want to remind him that, although he decided not to join a university faculty, he did have many students. John contributed to the way in which each of us formed and proceeded with our own research agendas. Moreover, those who contributed papers to this volume are just a sample of the many others influenced by his work.

This collaborative effort began several years ago and was intended to draw together economists who had worked with John at an earlier stage of their careers. Each contributor selected a research theme that reflects his or her own research in relationship to John's work. To provide readers with a sense of these themes as they were developed in John's research, we organized some of his key publications into five major areas, presented in the bibliographic profile at the end of the volume.

We also arranged to reprint one of John's seminal contributions, "Conservation Reconsidered" (see the appendix), because this paper has played a central role for resource and environmental economics as well as for the essays in this volume.

A number of individuals contributed to the volume. We would like to thank Resources for the Future and especially Robert W. Fri and John F. Ahearne, RFF's president and vice-president, respectively, for encouraging us with the project. Dorothy Sawicki did more than carefully edit the volume; she helped us capture the central themes of John's research through our own research chapters. While numerous secretaries and research assistants contributed to the individual chapters, a special acknowledgment is due to Sue Piontek for her role in helping assemble the complete first draft and all of the revisions of the manuscript.

Finally, our appreciation would not be complete if we failed to remember Shirley Krutilla's thoughtfulness and friendship. Weekends at the Krutillas' cabin in the Shenandoahs were special times that we won't forget.

August 1988

V. KERRY SMITH
University Distinguished Professor
Department of Economics and Business
North Carolina State University

PART I

BACKGROUND

THE INFLUENCE OF RESOURCE AND ENVIRONMENTAL PROBLEMS ON APPLIED WELFARE ECONOMICS
An Introductory Essay

V. Kerry Smith

This volume was prepared to honor John V. Krutilla, who has helped shape resource and environmental economics in a way and to a degree matched by few economists within their respective fields. His ideas and research accomplishments have been powerful forces in the development of the theory and the practice of economic analysis applied to resource and environmental problems. John's research career coincides closely with that of Resources for the Future (RFF), which was founded thirty-six years ago, in 1952. Indeed, of his thirty-six years as a resource economist, he has spent thirty-three as a member of the RFF research staff.

One objective of this introduction is to describe a part of John's influence, specifically, as it is evidenced in relation to the papers in this volume. As a consequence, this is probably a more technical essay than one might expect to find launching a festschrift. The explanation is simple. This is what John would probably prefer if he were given a choice. He would want us to consider his ideas, what they have meant for resource and environmental economics, and what they continue to mean for the research frontiers that lie ahead.

Nonetheless, if this essay communicated only those aspects of his work it would fall short of describing his impact in an important respect. For he also affected the way a large group of economists—including many of the contributors to this volume—select and complete their research.

Three important aspects of that influence on economists who were associated with him on various research projects or who otherwise followed his work should be mentioned: (1) his method of identifying and addressing a research problem; (2) his insistence on the importance of defining future research needs as part of the work of a project; and (3) that attractive quality about his ideas and research which engaged and motivated others in the field.

3

With regard to the first of these areas, it is significant that, while many economists look primarily to the profession and to professional journals for their ideas and stimulation, John has always gone directly to the set of resource policy problems for his. Simply stated, his strategy is this: identify a problem that matters; develop the theory, methods, and information necessary to address it; and be sure that the analysis and conclusions can be understood by those who might use them to improve resource allocation decisions.

The list of examples of this strategy from his own work is long. For the moment, the mention of three particularly important examples will suffice. The first is *Multiple Purpose River Development: Studies in Applied Economic Analysis*, which he wrote with Otto Eckstein (Krutilla and Eckstein, 1958). The purpose of the study, which was motivated by the Eisenhower administration's policy advocating a larger private-sector role in river basin development, was to consider the economic rationale for river basin development and the appropriate mix of private- and public-sector activities in that development. In carrying out that purpose, the work established the theoretical rationale for applied welfare economics and illustrated, in practical terms, how it might be used.

A second example is his work with Robert H. Haveman, *Unemployment, Idle Capacity, and the Evaluation of Public Expenditures: National and Regional Analyses* (Haveman and Krutilla, 1968). That study likewise set out to solve a particular problem and in the process laid the conceptual and practical foundation for another part of the structure of applied welfare economics as it is known today. The problem addressed in the study is this: the conditions for efficient resource allocation decisions assume that all resources are fully employed, and thus the real resource costs of one allocation can be readily gauged by the available alternative uses for those resources. The idea is fine for the classroom, but how can it work in cases where national or regional underemployment of resources may be more than reflections of short-term adjustments to a new full-employment economy? To solve this problem, the Haveman–Krutilla methodology introduced the concept of shadow prices for taking account of idle resources of manpower and industrial capacity, and it provided a practical means for estimating them. The shadow prices are based on the extent of employment and prospect for storage (thereby acknowledging the need to consider dynamic opportunity costs).

John is probably best known today for identifying the problems in making allocation decisions that involve unique natural environments and wildlife resources. The third example of his research strategy in operation is his pathbreaking article "Conservation Reconsidered" (Krutilla, 1967; reprinted in the appendix in this volume), in which he sought to identify a class of resources requiring a new set of analytical methods because of their special features. The features that he identified are these: in this class the allocation choices are irreversible; one of the uses of such a resource is unique (or at least has few substitutes); there is little or no prospect that its supply will increase over time as a result of expected technological change; individuals do not have to be active consumers to derive satisfaction from the presence of such resources; and frequently there are substantial uncertainties either in the factors affecting consumers' demands in the future or in the availability of supplies. The applications of this work were discussed somewhat later in his capstone volume, *The Economics of Natural Environments: Studies in the Valuation of Commodity and Amenity Resources* (Krutilla and Fisher, 1985, rev. ed.), published in 1975. Yet in his original article he structured a generic set of problems and identified some of the methodological advances that would be needed to deal with them.

The litany of examples could continue, but those given illustrate the first aspect of John's influence on his profession: through the example of his own work he taught a large number of resource economists about the process of planning and doing their research.

A second aspect of his influence reinforces the first: John has maintained that in answering a particular policy question, part of the researcher's task in preparing a detailed academic response to the original mandate is to define an agenda for the research that must logically follow. When the assumptions matter, he would counsel, find out why and then incorporate the revised conceptual and methodological formulations into the practical guidance necessary for future analyses of similar problems. The case studies in the books just cited illustrate this strategy, of which many papers in this volume are also the products.

A third way in which John's influence has made itself felt is through the infectious quality of his ideas and through his selection of an agenda of research problems that likewise captured the interest of a large number of young resource economists in the 1960s and 1970s and continue to do so today. Thus, his reach has extended well beyond his own considerable research accomplishments.

There seemed to be no better guide for preparing this introductory essay than John's own work. Thus, the title and organizing principles underlying this paper, as well as the rest of this volume, can be traced to his presidential address to the Association of Environmental and Resource Economists in 1980 (Krutilla, 1981). In it John provided a carefully woven blend of historical and methodological insights into the evolution of applied welfare economics. His essay offers the reader an insider's view of the intellectual environment at the time benefit–cost methods emerged as a tool for public policy analysis.

Leading theorists had concluded by the early 1950s that welfare economics had no practical relevance. In his Ph.D. thesis, *Welfare Economics and the Theory of the State* (the first edition having been published in 1952), Baumol (1967), for example, summed up the views of the time:

So long as we recognize the existence of particular types of interdependence in the results of the activities of our economic units, our analysis is likely to break down completely. . . . The fact that categories like "external economies" and "external dis-economies" remain largely empty economic boxes prevents any further application of welfare theory as it now stands. (Baumol, 1967, p. 205)

Juxtaposed with this view and continued theoretical laments about the lack of relevance of welfare economics was the "robust progress" that was being made in developing the practical means for dealing with externalities and judging the efficiency effects of specific, small-scale water resource projects. These efforts were clearly not what Baumol expected, according to the minor concession he offered in closing his rather gloomy judgment on the relevance of welfare economics: he noted, in effect, that it probably would not hurt if "practical economists" talked to policymakers.[1] Of course, as the preceding quotation implies, he did not hold out much hope that substantive advances in applying welfare economics would result from such consultations.

Writing over a decade later, Baumol dramatically changed his appraisal. In a lengthy introductory essay to the second edition of his thesis (Baumol, 1967), he clearly acknowledged that benefit–cost analysts of the 1950s and 1960s were able to demonstrate

how specific governmental projects can be evaluated in a manner which accords at least roughly with the prescriptions of welfare analysis. . . . [T]hese writers have also breathed life into many of our welfare constructs even though they have been forced to treat some of them rather roughly at times [emphasis added]. (Baumol, 1967, p. 23)

In the discussion that follows, I argue that practitioners of applied welfare economics over the past three decades have not simply *used* the "well-oiled" theoretical machinery of welfare economics, as Baumol seems to imply. They have advanced it![2] Moreover, the issues raised in evaluations of resource and environmental problems have themselves been at the leading edge of this process.[3] Certainly John Krutilla's work has been a central element in this development throughout the three decades since the publication of his first major work, *Multiple Purpose River Development*, in 1958.

Like applied welfare economics, part I of this introduction develops its case using both the theoretical and the applied literatures in considering the following issues that have been influenced in important ways by Krutilla's work: (1) efficient allocation decisions and policy design when externalities are present, (2) the valuation of nonmarketed resources, (3) the selection of a discount rate for public investment decisions, and (4) public management of environmental resources. While these four issues are not entirely independent of one another, they are described separately in this paper for the purpose of identifying some of their primary lines of research. These particular issues were selected because they are especially relevant to the other essays in this volume. Rather than summarizing the individual papers, part I of this introduction discusses the relationship of each to these issues or to the broader inferences one might draw from Krutilla's research.

Part II highlights areas in which future advances in the theory of applied welfare economics seem likely, based on the shape of today's resource and environmental problems.

PART I
APPLIED WELFARE ECONOMICS: THEORY AND PRACTICE

Applied welfare economics provides the principles for making investment and management decisions that result in efficient resource allocations; these principles apply in cases where the inputs to or outputs from production activities require some form of intervention or supervision by the public sector.[4] Therefore, applied welfare economics includes what is conventionally referred to as benefit–cost analysis, as well as the analyses associated with defining the operating and access conditions for production processes or resources under public control or oversight.

A simple definition of efficiency is helpful in explaining the two types of decisions (investment and management) and in organizing the discussion to follow. It suggests that resources should be allocated to their highest-valued uses. For Western, free-market economies, the values used for making such judgments are assumed to be those of private consumers. Thus, while acknowledging that the distribution of income (and wealth) is important to a tangible realization of consumer sovereignty in the form of the monetary values placed on goods and services, the analysis generally proceeds by accepting a particular distribution of income (usually the status quo) and gauging the efficiency implications of a specified array of actions.

An examination of the distributional implications of the actions is often supplementary information, developed independently of any judgments about efficiency. Much of the literature in the 1960s viewed the focus on efficiency as being a serious limitation in the application of benefit–cost analysis in the United States. Maass (1966), for example, contended that the true objective of most public spending was redistribution of income, not improving the efficiency in the existing resource allocation. The statement by Maass and related work in this area require additional comment.

In the 1950s and 1960s most benefit–cost analyses were evaluations of public spending projects. Consequently, discussions of the implications of accepting the Kaldor–Hicks criterion for potential compensation of losers (where public actions impose both gains and losses) were routinely arguments over equity goals and about whether it was appropriate to specify in advance a "trade-off ratio" or set of weights for efficiency and equity goals.[5] Implicitly, the gains and losses were viewed as being associated with specific income groups. With the perspective that only an "insider" can offer, Allen V. Kneese, in the next paper in this volume, provides additional insight into the evolution of research on the most important component of the public investment projects of those early years—the water projects.

Today, the focus of the distributional analysis differs depending on the type of public action being evaluated; regulations imposed on private-sector activities are now analyzed at least as frequently as are public spending projects. For the latter, the incidence of distributional effects by region, by sector, or by some other aggregative criterion is often reasonably clear from the definition of the particular project, so the focus is on the groups in the region that will gain. By contrast, in the case of regulatory actions, the legislation defining the programs often deliberately impedes a priori appraisal of regional effects (see Zeckhauser [1981]), so the focus of distributional analyses is the region or sector of impact. Neither approach should be interpreted as an effort to gauge the shortcomings of the Kaldor–Hicks standard relative to some theoretical ideal. Instead, they identify impacts that have political significance for the respective types of decision.

Investment decisions and management decisions use quite different methods of implementing a practical version of the efficiency criteria (i.e., one based on the Kaldor–Hicks concepts). As a result, the two types of decision have different connections to welfare theory as well as different ways of advancing it. As a rule, benefit–cost analysis involves an evaluation of whether the departure from the existing resource allocation implied by a specific (usually small) project[6] would yield a greater increment to aggregate benefits than to aggregate costs. It signals opportunities for improvement. (Otherwise, if the existing resource allocation were efficient, there would not be positive net benefits.[7]) Thus, the connection of an investment decision to a theoretical description of the conditions for an efficient resource allocation (whether static or dynamic) is less direct than in the case of a management decision. As Bradford (1970) has argued, positive net benefits in the aggregate describe conditions sufficient to improve the allocation of resources from the status quo and that is all. They do not assure the optimality that the relevant marginal conditions would guarantee for well-behaved benefit and cost functions.

By contrast, the use of efficiency criteria in managing an existing set of resources (or production activities) is often treated as an attempt to implement the marginal criteria implied by static efficiency. For example, how should the conditions for access to public grazing lands be defined when private interests (in supporting cattle) are in competition with the mixed public–private purpose of sustaining a particular wildlife species that provides consumptive and non-consumptive values? The nature of the problem drives the methods that are developed to approximate the relevant marginal conditions for efficiency.

Of course, the two problems and how they are treated are not independent. Indeed, more than twenty years ago in his evaluation of the prospects for achieving efficiency goals in public policies affecting water resources, Krutilla

(1966) was among the first to provide an explicit discussion of this point. He observed:

We have on the one hand dicta that the scale of a facility should be extended to the point at which incremental benefit equals incremental cost. However, unless user charges are levied to cover the cost of providing the marginal unit of output, there will develop excess demand represented by those users who would not find incentive to use the services of the facility if charges appropriate to the design criteria were imposed, but who will make use of the facility at zero price. If such use by any beneficiary at capacity output adversely affects the utility of any other, the design criteria result in a project inappropriately sized relative to realizing the benefits estimated for the purposes of design. . . . [D]esign criteria as presently advanced relate to the correct design for an irrelevant situation where reimbursement policies are at variance with design criteria [emphasis added]. (Krutilla, 1966, p. 72)

Moreover, the issue is not simply one of defining access conditions which are consistent with the values that a project's aggregate benefit estimates imply consumers will derive. Rather, a correct evaluation of a public investment involves all of the assumptions about how the resources provided by the project would be used. More generally, the definition of practical efficiency criteria for public investment or management decisions requires an authentic description of the production and consumption processes.[8] For wilderness recreation, for example, this can amount to a recognition of how management practices affect the quality of the recreation provided by specific sites, the corresponding implications for consumers' valuations of trips with alternative quality levels, and, finally, the resulting increments to aggregate net benefits from changes in the stock of the areas under different management regimes. The Bowes–Krutilla (1985) analysis of the implications of multiple use criteria for the management of public lands implies that the timber-harvesting policies must be made recognizing the interdependencies in production between stands of trees of various ages, wildlife habitat, watershed maintenance, and recreation.

Ideally the two types of decision—management and investment—would be solved simultaneously. While a concurrent solution represents an ideal, it is not, at present, relevant to practice. Consequently, the following discussion of four issues in applied welfare economics tends to emphasize the policy task (whether allocation or management) with the most significant effects in relation to the issue under discussion. This is not to imply the other task would be immune to influence but is merely to confine the discussion to manageable length.

Externalities

External effects—by-products of one economic agent's actions that affect the production or consumption activities of others without a mechanism that adequately compensates for these impacts—have been among the clearest theoretical justifications for public intervention. Nonetheless, as numerous writers (for example, Haveman [1970]; Krutilla [1981]; Page [1973]) have acknowledged, the early welfare economics literature questioned their potential importance because of a failure to "discover" concrete examples of the phenomenon.

While resource and environmental problems offer a ready solution to this dilemma with examples ranging from the familiar air and water pollutants to less-fully-understood hazardous wastes, the impact of these problems on theory has actually been greater than simply to provide actual cases in point. For example, without specific cases to consider, the theoretical analyses of resource management in the presence of externalities tended to treat each agent as an equal contributor to the potential external effect. Thus, appropriately defined, uniform taxes per unit of contribution on the generators (if the effects were detrimental) offered the efficient policy instrument. Indeed, even if the marginal damages associated with the actions could not be observed, iterative adjustment in such uniform prices for the externalities seemed a convenient way to approach the theoretical ideal (see Bohm and Russell [1985] for a summary).

The reality of environmental problems quickly changed that. Once it is acknowledged that agents can and do contribute in different degrees to the external effects and that this is an essential dimension of any economic analysis of this policy problem, then it is also clear that the optimal taxes (charges) for efficient resource allocations require tailoring to reflect source-specific costs. This type of insight was apparent in Krutilla and Eckstein (1958, table 2). Here the differentials of each hydroelectric project's contribution to the power output are direct and indirect—impoundment at one location has implications for the performance of other units downstream. Moreover, the connections extend to different outputs in a manner analogous to the pollution cases. The managed river system provides navigation and flood control. Attempts to ignore these outputs or to separate them from the power system can, depending on the physical features of the river system, lead to the production of significant externalities.

Thus, in a genuine sense, such examples—beginning with the water resources cases and extending to virtually all real-world pollution problems— have motivated the design and theoretical analysis of new forms of policy instruments (see Bohm and Russell [1985]; Tietenberg [1986]).

Equally important, resource and environmental problems revealed a new class of externalities—intertemporal and intergenerational externalities. Certainly, when we define the parties involved in interactions where compensation is not possible, the external effects can be treated as extending over time as well as distance. For example, increased recognition of the presence of stable, accumulating pollutants such as carbon dioxide and chlorofluorocarbons in the atmosphere suggests a need to be aware of the special implications that different types of intertemporal externalities have for efficient resource allocations (see Cropper [1976]; d'Arge and Kogiku [1973]; and Smith and Krutilla [1979] for further discussion). Indeed, before the consideration of examples, it was less obvious how important the features of the externalities and our state of knowledge about them are to our ability to interpret the dynamic conditions for efficient resource allocations.

In general, the resource examples served to identify a class of intertemporal externalities that arises because certain allocation decisions are irreversible. Insofar as there is imperfect knowledge of future events, especially those directly related to the consequences of a particular allocation, then decisions that entail a permanent commitment of resources can be viewed as imposing an intertemporal externality on future generations. Krutilla (1967) was the first to identify this class of problems, relating them with preservation-versus-development decisions that might affect unique natural environments or the habitat of an endangered species. His classic essay "Conservation Reconsidered" (see appendix in this volume), published two decades ago, launched a subfield within resource and environmental economics and, more generally, a new view of public investment decisions. Irreversibilities were clearly established as important for the application of benefit–cost criteria in decisions involving unique natural environments.

As one might expect, the problems posed by irreversibilities have been characterized in many ways. The original approach described the decisions in an optimal control framework (Fisher, Krutilla, and Cicchetti, 1972). The features of the problem led to the conclusion that a socially optimal investment profile required either immediate investment in the development option or preservation in perpetuity.[9] Subsequent studies (notably, Fisher [1981] and Porter [1982]) have used somewhat simpler frameworks to illustrate the key elements of the problem: the assumed asymmetry in the changes in benefits from the alternative uses of the resources involved over time; the timing of the investment decisions (and their relative indivisibility); the treatment of discount rates; and the character of the uncertainty, including the way in which it is assumed to be resolved.

The papers in this volume by Fisher and by Brown and Swierzbinski discuss issues that emerge from these early insights. Fisher evaluates the conventional economic analysis of species extinction. His paper returns to two problems identified as being important to the Krutilla thesis that the conventional practices of benefit–cost analysis were systematically biased in favor of the development of unique natural environments because these methods failed to account properly for the special features of the resources in question and for the consequences of development decisions.

Fisher begins his paper by describing how in the presence of uncertainty it is possible to overinvest in development even when one seemingly recognizes the uncertainty in the process. His result follows directly from a value-of-information argument. If, after the uncertainty is resolved and the net benefits from alternative allocations are known, it is impossible to change a subset of the possible decisions (i.e., development is irreversible), then we cannot use anything we learn that resolves all or part of the uncertainty. That is, one needs to take into account the conditional value of information provided by avoiding irreversible actions. In this case the values from the information expected to become available with the passage of time (or the investment of resources) can be realized *only* if the irreversible actions have *not* been taken. The message is clear: socially optimal behavior requires both proper treatment of the uncertainty and recognition of the feasible outcomes under alternative sets of decisions. It is not sufficient to use expected values, *even if the appropriate social attitude toward risk is one of risk neutrality*.

Fisher closes his paper by addressing the role of such externalities in the case of allocation decisions without uncertainty. As with his first argument, he uses endangered species problems for his examples; nonetheless, the principles involved are more general. It is often asserted that exploitation of a renewable resource can proceed efficiently to extinction[10] and thus that preservation must be justified on noneconomic grounds. This need not be the case, Fisher argues. The unharvested stock of the resource can be a store of value—genetic information, for example. Recognition of the loss of these values changes the economic conclusion that we can take comfort in knowing we "efficiently eliminated" a species. Indeed, extinction is not necessarily the efficient outcome. The Brown–Swierzbinski paper also considers this second type of effect (that the unharvested stock may have independent value) and extends the Fisher argument by highlighting how the definition of the value function used for species as a continuing source of information can have far-reaching implications.

Once it is recognized that the stock of natural endowments may well be a nonpriced, essential input to the production of research and development (R&D) in the private sector, another of the themes of resource and environmental problems emerges. That is, in order to capture the linkages that explain some of the importance of natural and environmental resources for economic activities, policy evaluations must recognize the technical relationships governing the ways in which resources are combined in production or are actually used in consumption. These technical relationships substitute for the signals we can expect with marketed resources. Market transactions provide a basis for determining the roles of priced resources in production and consumption activities, but the services of nonpriced environmental resources would go unnoticed under this criterion. Thus, one approach to determining the contributions of the latter involves looking more closely at the explicit details of the activities of production and consumption. As is described by Kneese (in this volume), this point was most clearly developed and illustrated within the Water Quality Program at Resources for the Future.

This message is relevant for Brown and Swierzbinski's analysis because they find that the effects of externalities and market structure in one set of activities can influence seemingly unrelated policy issues in another context—namely, the valuation of the stock of genetic resources. To develop this argument, the authors first consider the positive externalities that arise as firms produce new knowledge in the form of R&D. To the extent that these external effects can be controlled so as to appear appropriated, private actions of the firms will lead to underinvestment in R&D. Their argument, an extension of earlier work by Spence (1984), demonstrates how important the assumed mechanisms for the externalities (i.e., symmetric versus asymmetric) and for the product market structure are to the outcomes. The more the results of R&D activities are appropriable and symmetric, the greater the efficiency losses. The result is underinvestment (because of failure to account for the positive external effects that would accompany open access to the information generated).

However, the implications of Brown and Swierzbinski's analysis extend beyond this enhancement of Spence's work on the interactions between optimal R&D levels and industry structure. Their conclusions have implications for the valuation of genetic capital as well. Within their framework they assume that the value of genetic capital derives from its contribution as an input to the Brown and Swierzbinski production of knowledge. Any factors tending to reduce the level of R&D activity will, in turn, affect the value society places on these genetic resources—in general, leading to undervaluation. External effects, such as institutionally induced market distortions, have general equilibrium consequences. Moreover, the causality flows both ways, not only from the organization and regulation of industry to the implicit (derived) valuation of genetic resources but also in the opposite direction. Policies that lead to inefficient exploitation, to overexploitation, of these genetic resources (i.e., to irreversible depletion of the stock) may have serious long-term consequences by affecting the production possibilities in future R&D efforts. These are precisely the types of uncertainties that the first argument in Fisher's paper addresses.

Consumers' Valuations of Nonmarketed Resources

Once it is seen that problems associated with the allocation and management of natural and environmental resources provide numerous examples of external effects and nonmarketed commodities, it seems reasonable to take the next step: to consider whether these problems also played a role in the theory and the practice of determining how such resources were to be valued. Here, too, the answer seems to be that they did have an important role. The practical tasks required for analyzing resource problems not only provide grist for refinement of the concepts in the theorist's mill, but also led to the identification of issues not previously appreciated or incorporated in valuation theory.

Of course, to suggest that the process was unidirectional, with the applications always promoting new theory, would be misleading. As in the case of the refinements and extensions to theoretical analysis of externalities, there seems to be a continuing two-way flow.[11] While each observer of the developments since Krutilla and Eckstein (1958) could be expected to highlight a somewhat different subset of the issues in obtaining measures for consumers' valuations of nonmarketed resources, four developments would certainly appear prominently in many of these lists. They include (1) the identification of a clear distinction between use and nonuse benefits, (2) the recognition of the role of behavioral responses to uncertainty (both for the consumer and for the policymaker) in the definition of valuation concepts, (3) the treatment and valuation of quality-related attributes of environmental resources, and (4) the development of direct and indirect methods for estimating individuals' valuations for changes in the amount or in the characteristics of nonmarketed resources.

Use and Nonuse Values

While there was discussion of the motives for valuing natural endowments and endangered species in the writings of Ciriacy-Wantrup (1952), Krutilla's "Conservation Reconsidered," published in 1967, is justly credited with being the first comprehensive recognition of the constituent elements of nonuse values. In addition, it identified the importance of consumers' responses to uncertainty in the valuation of resources by pointing out the need to consider the special role that the Weisbrod (1964) notion of option value might play for unique natural and environmental resources.

Krutilla's short, illustrative descriptions of these concepts fueled extensive subsequent work. While it is not possible to summarize all of it here, a few dimensions of the use–nonuse distinction in the task of developing estimates of consumers' valuations should be mentioned. First, the definition of *existence value*—the value generated because the presence of a particular resource (or improvements to it) contributes to the utility of some individuals even though they do not make specific use of that resource—implies a fundamental redefinition of what is meant by consumption. That is, if it is possible for a single individual to have separate values for the use and the existence of a resource, then the services provided by the resource generating these values must enter the utility function in distinct ways.[12] Moreover, it is necessary to reconsider the conventional notion that the act of consumption is what leads to utility and that it involves a using up of the commodity within the relevant time period (i.e., the time horizon defined within the model).

It may be important to any conceptual justifications for existence value to determine whether the ultimate reason for direct use and existence of a resource serving as distinct contributions to utility arises from an individual's altruistic motives (Randall, 1985; Randall and Stoll, 1983) or from indirect forms of nonconsumptive use (that are hard to detect). But the reasons do not change the implied departure from conventional practice in modeling the ways in which individuals use goods and services. This was indirectly recognized in McConnell (1983) and Freeman (1985) and more specifically discussed in Smith (1985).

Equally important, the existence of such motives has important implications for the use of Mäler's (1974) weak complementarity restrictions in any effort to estimate the demand for a nonmarketed commodity by means of indirect methods. These restrictions preclude any form of existence values by suggesting that an observable action, such as a positive demand for a marketed private good, is required for an individual to have a nonzero valuation of the nonmarketed environmental resource whose demand is to be inferred. By restricting preferences so that the nonmarketed commodity's consumption is linked to the marketed good, weak complementarity allows the analyst to use information about the latter to recover demand for the former. Existence values imply that individuals have value for a resource independent of any form of use, and these values would therefore be precluded if this approach were used for benefit estimation. Preference structures consistent with these values will not satisfy Mäler's restrictions.

Uncertainty and Valuation

In his initial discussion of option value, Krutilla (1967) treated it as part of the nonuse values that must be considered for unique natural environments or species. Subsequent literature has concluded that his implicit taxonomy is confusing (see, for example, Boyle and Bishop [1987]; Hanemann [1984]; Smith [1987]). This new work argues that the task of valuation under uncertainty implies a change in the standard used to define well-being. That is, once uncertainty is incorporated into a model of consumer decisions, the implied behavioral responses not only affect the specific monetary values derived for policies but can affect *the valuation concepts used to describe these monetary values*. This potentially significant, if subtle point indicates the importance of treating von Neumann–Morgenstern and conventional utility functions as offering equivalent measures of an individual's well-being. The former offers an ordinal scale of well-being *in advance* of realization of the events that are uncertain to the individual; the latter (whether ordinal or cardinal) is an index of well-being *after* consumption. When policy actions must be valued before the uncertainty is resolved, an *ex ante* approach that underlies the von Neumann–Morgenstern utility concept uses values based on preferences and beliefs before that resolution, while the *ex post* is based on preferences after the uncertainty is resolved.

As Krutilla anticipated, the potential importance of the distinction in the valuation concepts implied by each standard depends on just how unique the good or service is to each individual (Cook and Graham, 1977; Plummer and

Hartman, 1986; Smith, 1984a). This is a recurring theme in his and other discussions of how to gauge the types of resources and individuals whose monetary values are likely to be greatly affected by the welfare concept used (i.e., constant utility or constant expected utility). For example, Fisher and Krutilla (1985) have recently repeated and distilled these arguments for the case of the preservation of natural environments noting that

there is a legitimate question whether undisturbed natural environments should be allocated to extractive industrial activities on the supposition that they can be restored eventually to provide replicas of the original that would satisfy the recreational interests of only the less discriminating clientele. The matter turns on the importance of authenticity as an attribute of the recreational experience. . . . (Fisher and Krutilla, 1985, p. 177)

Those with strong preferences for authenticity tend to recognize few substitutes, and, as a result, have highly inelastic "demands" for the services of a specific resource. When the policy issue involves loss of a resource that is considered unique to these consumers, the feature of individual preferences that the Krutilla–Fisher argument describes can be represented by differences in the marginal utility of income depending on the availability of the resource.[13] Large differences can be expected to affect not only the size of option value (as the examples provided by Krutilla implied) but also all measures of an individual's risk aversion in such cases (see Karni [1983, 1987]).

This concept, to be distinguished from what Fisher (in this volume) labels as an option value, requires concave preference functions that are distinguished by state of the world. It results from attempts to define monetary measures of changes in individuals' well-being once it is recognized that behavior changes with uncertainty. By contrast, Fisher's framework abstracts from such changes in behavior. His concept is based on an implicit model of a policymaker's actual behavior in comparison with ideal behavior (i.e., theoretically consistent decisions) when faced with specific types of uncertainty.

This distinction between monetary values measured with different concepts of well-being for the individual and for the policymaker also arose as resource economists worked out ways of implementing these concepts in applied welfare economics. As Kneese's paper (in this volume) on the evolution of water resources research suggests, such distinctions in valuation concepts are likely to grow in importance with attempts to address the problems of *ex ante* regulation versus *ex post* cleanup of groundwater. The prospect of significant contamination of groundwater with toxic chemicals and of corresponding long-term health risks for large segments of the population indicates that these conceptual and empirical issues are likely to be a part of the next decade's water research.

Valuing the Quality of Environmental Resources

Prior to the Fisher–Krutilla (1972) analysis of the effects of congestion on wilderness recreation, economists had tended to treat congestion within a cost-based framework. For example, congestion on highways or in service-related activities was assumed to lead to increased waiting (or servicing) time. It was natural, therefore, to model the effects of congestion in terms of the increases to the marginal costs of completing the trip or obtaining the services (see Edelson [1971]; Johnson [1962]; or Walters [1961] as examples). As described below, the Fisher–Krutilla (1972) paper and the research that followed from it are excellent examples of the systematic Krutilla approach to a research issue.

For purposes of analyzing resource problems, waiting time has no meaning in describing the important features of a wilderness recreation trip. Consequently, it was necessary to consider how increased use of an area affected the wilderness recreation trips of users. To do so, Krutilla examined the behavioral research of geographers, sociologists, and those social scientists familiar with the characteristics of wilderness experiences. He wanted to describe the attributes of such trips that were important to the recreationist's enjoyment. The pioneering work of Lucas (1964) and Stankey (1971) provided the answers. Wilderness enthusiasts wanted to be able to determine whose company they would be in during a trip. Solitude was their goal, and interruptions of that state, measured by encounters with recreationists who were not in their own party, reduced the quality of their experience.

Congestion viewed in this context should be modeled as a negative influence to these recreationists' valuations of their experiences. In fact, the conventional situation, using a cost-based characterization, is simply a special case of the framework proposed by Fisher and Krutilla.

Nonetheless, there remained the important issue of judging whether the treatment of the effects of congestion on valuations of recreational experiences would influence management decisions. Moreover, if it did, how could the theory be made operational? The answers followed as part of the research agenda of the Natural Environments Program at Resources for the Future. While not directly connected with the program at the time of completing their paper on "Congestion, Quality Deterioration and Heterogeneous Tastes," Freeman and Haveman (1977) were motivated by earlier program involvement to consider the effects of heterogeneous tastes and the assumed patterns of access to a recreational facility for the definition of the optimal capacity of a resource. This work established on theoretical grounds that the issues raised in the Fisher–Krutilla (1972) paper were potentially important. Other research efforts were directed toward estimating individuals' willingness to pay for wilderness, taking account of the effects of congestion as they had been found

to be related to solitude (Cicchetti and Smith, 1976), and models were developed for estimating how the aggregate patterns of use led to congestion
(Shechter and Lucas, 1978; Smith and Krutilla, 1976).

While the results were specific to the case of wilderness recreation, this
work offered a preview of an important part of today's research. (It also
anticipated some of the issues that would arise in club theory in defining
optimal capacity with heterogeneous members; see Cornes and Sandler [1986].)
Subsequent research on the effects of water quality for recreation or on the
characteristics of hunting areas as they affect recreationists' valuation of hunting
trips and more general studies of the valuation of quality have been influenced
by this work. It is particularly significant for recognizing the importance of
identifying how individuals perceive the quality of the commodities they demand and how resource allocation decisions could be linked to factors that
influence those perceptions.

Direct Versus Indirect Benefit Estimation

The last area on the issue of consumers' valuation of nonmarketed resources
to be discussed here involves recent attempts at developing valuation models
capable of responding to the requirements imposed by policy problems while
maintaining theoretically consistent welfare measures. Led by publication of
The Benefits of Environmental Improvement: Theory and Practice (Freeman, 1979),
in which the author synthesized his work and the literature to that point, and
driven by the need and budgets of the Environmental Protection Agency (EPA),
a substantial portion of recent research in resource and environmental economics has focused on methods for valuing features of environmental resources.
As Kneese (in this volume) indicates, this work was stimulated by the growing
recognition of the costs of regulatory programs intended to improve environmental quality during a period when the overall performance of the economy
was not good. A program of phased implementation had been built into the
major environmental legislation of the 1970s, thereby delaying the actual impact
of these regulatory initiatives.[14]

Wilman (in this volume) describes the theoretical implications of using three
possible methods for valuing the attributes of recreational resources. This work
is one of the results of Krutilla's current program on the implementation of
the multiple use and sustained-yield directives in the legislation governing the
management of public lands (see Bowes and Krutilla [in press]). This research
program was based on the clear recognition that the management practices
adopted for public lands affect not only the capacity of these lands to provide
recreation but, equally important, the quality of the recreational activities as
well.[15]

Wilman's model was developed for use in estimating the value of the quality
features of hunting recreation (see Wilman [1986]). She considers in some
detail the modeling assumptions that influence the use of the conventional

travel cost, hedonic travel cost, and simple repackaging models for this type of recreation.[16] The selection of a model to estimate these valuations requires that we explain how the characteristics to be valued influence behavior. They can play a role in the decision to engage in the activity, in the selection of areas to hunt, and in the number of times the activity is undertaken in any particular period. To describe each effect adequately, it must be specified how individuals perceive these characteristics as influencing the supply options they face in making these choices. Wilman's analysis describes the problems in implementing three possible models for valuing hunting experiences.

Her recognition of the interaction between the actual features of the activities to be modeled, the characteristics of the resources to be valued, and the limitations imposed by the information available for implementation illustrates why theory and practice must be jointly considered in the use of indirect methods for estimating consumers' values of nonmarketed resources.

The Krutilla research program also offered some of the first examples of the use of survey techniques—the contingent valuation method—to estimate individuals' values for the attributes of environmental resources. My work with Cicchetti referred to above (Cicchetti and Smith, 1976) was part of the Natural Environments Program. The Fisher–Krutilla (1972) framework had implied that disruptions to solitude should reduce wilderness recreationists' willingness to pay for trips. To evaluate the empirical support for this hypothesis, we mailed a questionnaire in which hypothetical experiences were described to individuals who had used the Spanish Peaks Area in Montana in a season preceding the date of our survey. The results not only confirmed the Fisher–Krutilla hypothesis but illustrated the value of using experimental design as part of a contingent valuation survey.

Our understanding of survey methods has advanced dramatically since that time, largely as a result of the work of Brookshire and coauthors (1982), Cummings, Brookshire, and Schulze (1986), and Mitchell and Carson (1988). Krutilla's program was one of the several early efforts that supported the use of survey techniques for estimating the values of attributes of environmental resources when the data requirements implied by the models that Wilman describes could not be met.

The Discount Rate

Public investment decisions involve projects that extend over long periods of time. As a consequence, benefit–cost analyses have routinely been cast in present-value terms, and the economic issues associated with the selection of a discount rate for these calculations have always played a prominent role in resource economics.

Indeed, one of the contributions identified with the Krutilla and Eckstein (1958) volume is the authors' proposal for calculating an appropriate discount rate. Their analysis argues in favor of an opportunity cost criterion in selecting a discount rate, because a public investment should be viewed as "a loan by society to itself in order to build certain physical investments" (p. 89). This does not imply that there is a single opportunity cost. Nor does it suggest that the opportunity cost would be the rate of return on the forgone marginal private investment. In the Krutilla–Eckstein argument such a position would ignore the realities of the public investment decision. There exist at any point in time widely differing rates. At issue in selecting a rate under their approach is determining the actual sources of the capital used to finance the project involved. That is, "a sector-by-sector approach, assuming a specific incidence of marginal taxation, is far more trustworthy *because it corresponds to the actual conditions under which public capital is raised*" [emphasis added] (Krutilla and Eckstein, 1958, p. 91).

Many subfields in economics have analyzed the discount rate issue. Resource and environmental applications are, in this case, one of many stimuli to the evolution of these arguments.[17] Some of these developments that came from Krutilla's program of research after his first contribution in *Multiple Purpose River Development* should be noted. These three are the most important:

1. The adjustment in discount rates to reflect the special features of natural resources,
2. The appropriate treatment of risk in the selection of a discount rate, and
3. The question of intergenerational equity and the rate of discount.

The last of these is often associated with Page's work at Resources for the Future (see Page [1977a, 1977b]; Ferejohn and Page [1978]). (As discussed in more detail below, Page's paper in this volume offers a new focus on the issue of intergenerational equity, arguing that it is more than simply a decision to select a specific value for the discount rate.) The other two contributions can be viewed as having added new complexities to the selection of a discount rate. The first responded to suggestions that we should tinker with the "appropriate" discount rate to reflect the special problems or features of natural resources. These features might include progressive changes in the real values of some of the services provided by these resources in a preserved state or the potential implications of irreversibility resulting from a public investment decision. In both cases, the position developed in resource economics has been to reject tinkering in favor of explicitly incorporating the changes in real values in the calculations or modifying the decision rule itself for irreversibilities. (This argument is developed in detail in Fisher and Krutilla [1975].)

The second contribution is also theoretical. Fisher (1973) recognized that the Arrow–Lind (1970) suggestion for the treatment of uncertainty in public investment decisions may well be inappropriate for many natural and environmental resources. That is, Arrow and Lind's conclusion favored a risk-neutral position for the public sector. It relied on the spreading of an ap-

proximately constant aggregate risk over a large number of taxpayer-benefi-
ciaries so that each one's share could be treated as being arbitrarily small.
Fisher's argument is direct: in cases where the outputs provided involve public
goods or where the allocation decisions imply significant irreversibilities, the
constancy of the aggregate risk is not assured. Thus, the diminishing risk per
beneficiary would not arise.

The conclusions reached with regard to the first two issues above are
consistent with the argument that the role for any feature of a resource or of
the decision context (e.g., uncertainty) should be explicitly reflected in the
benefits and costs and not buried in ad hoc discount rate adjustments.

Page's paper (in this volume) is in the same spirit, but in many respects it
calls for more fundamental changes in the terms of the debate over intergenerational equity. In Page's view the question is not one of selecting a discount
rate but rather of selecting a set of axioms that describe an appropriate
intergenerational social choice rule. These axioms, after all, provide our basis
for judging what is "fair." By comparing the Koopmans (1972) axiom set and
its support for discounting with the original Arrow (1963) axioms, both under
conditions of an infinite generational time horizon, he effectively illustrates
how the analysis is limited by focusing on the selection of a value for the
discount rate and thereby accepting a present-value criterion.

Resource-related problems clearly motivate consideration of the issues he
raises. Accumulating atmospheric carbon dioxide, disposing of commercial
nuclear waste, and husbanding the natural resource base are all problems that
stretch the conceptual basis for benefit–cost analysis well beyond the bounds
for which it was intended—a single generation borrowing from itself. Resource
economists have been among those leading the discussion of how to adapt
theory to fit the needs of policymaking in these areas.

Public Management of Environmental Resources

In the United States, water and land resources are involved in a large portion
of the resource management activities undertaken by the federal government.
As described earlier, public projects are mechanisms through which society
develops the public capital necessary to achieve some objective not met by
private markets. Benefit–cost analysis is the means for evaluating whether any
public investment designed to provide that capital would be warranted in terms
of crude efficiency judgment. Once undertaken, these projects provide public
hydroelectric-generating facilities, dams for flood control, lakes for recreation,
canals for transportation, irrigation networks, reserved public forest lands and
recreation areas, and an array of other facilities. As a rule, the legislation
authorizing such investments assigns management responsibilities for them to
an existing government agency (or it establishes a new agency for that purpose).

The legislation also defines the managerial objectives and constraints governing the administration of the resources involved.

While this description makes the whole process of management seem reasonably straightforward, it is not. Being the result of compromise, legislation often avoids specifics. It is left to the administrative rule-making process to add relevant details and to resolve conflicts by attempting to interpret the intent of the lawmakers, thereby providing a workable basis for agency action. Since the purpose of public intervention is usually to mitigate some private-market failure, efficiency is typically an important component of the operating principles of management agencies.

Because of the dominant role of natural and environmental resources in the set of activities managed by the federal government, resource economics has likewise played an especially important role in defining the conceptual principles to govern these operations and in highlighting the problems associated with existing practices.[18] Five papers in this volume (those by Freeman and Norris, Hyde and Daniels, Howe, Davis, and Cicchetti and Haveman) illustrate the range of theoretical and empirical problems raised by attempts to describe the principles warranted for efficient management of natural resources as well as to evaluate the practices in use by resource agencies. All five deal with themes that can be traced to Krutilla's work.

Two relate directly to the problems Krutilla (1966) identified with the efficiency of water resource agencies. Freeman and Norris discuss the theoretical relationship between the literature on optimal pricing in the presence of budget constraints and the cost allocation and reimbursement policies for both water and land use programs. The paper by Hyde and Daniels discusses the problems of implementing the multiple use mandates for managing public forest lands in the presence of constraints.

Krutilla (1966) is cited earlier as describing the importance of the interaction between the principles used for managing public resources and the criteria proposed for public investment decisions. In the same article he highlighted the need to allocate costs and to design reimbursement policies in a framework consistent with efficiency criteria: "In the absence of a pricing policy consistent with efficiency criteria, either an adulterated quality of service will be provided or a continuously existing over-capacity will prevail" (Krutilla, 1966, p. 72).

Reimbursement policies are usually said to involve two steps. The first requires resolution of a cost allocation problem. That is, after the separable costs of a project's (or an ongoing activity's) outputs are allocated among those outputs, the nonseparable or joint costs must be allocated. The second step calls for a policy decision to determine the amount of these costs that is to be reimbursed by the project beneficiaries. In other words, pricing policies must be established directly or indirectly.[19]

That second step remains a persistent problem in public water resource projects. It was identified as such by Krutilla twenty years ago and again by

Young and Haveman (1985) in their review of the literature on the economics of water resources. Howe (in this volume) provides clear-cut evidence of the extent of the problem. For example, in six federal irrigation projects that he discusses, the prices range from about one-fifth to nearly one-twenty-fifth of the estimated full cost of providing the irrigation water. This, together with the other information that he summarizes, presents a clear record of failure to efficiently deliver *and ration access* to public water projects. Federal water resource policy emerges as a continuous effort to separate the beneficiaries of water projects from those who pay the costs.[20]

Howe's conclusion, while discouraging, reflects another component of the research using applied welfare economics to evaluate policy: it has identified, sometimes indirectly, the political economy of resource problems. That is, evaluations of policies in the environmental area invariably demonstrate the importance of the institutions governing the policymaking process. Indeed, this theme clearly emerges from the account of the process given by Davis (in this volume). While Davis's paper is a tale of implemention of the Endangered Species Act with respect to the snail darter, it is also the story of a water project—the Tellico project. To appreciate the discrepancy between a set of practical guidelines for efficient management and the realities that the political process produces, contrast Davis's lessons from the politics of the Tellico decision with Howe's admonitions for improvements in the management and investment processes. Davis tells us that, according to the political rules:

1. Deals must be respected. Authorization of a project, right or wrong, assigns "property rights" to one (or more) congressional district(s) and therefore to some set of beneficiaries.
2. Mistakes are not be admitted. Specifically, to acknowledge waste is far worse than wasting more: i.e., all sunk costs are viewed as real or political opportunity costs.
3. The decision becomes irreversible and efficiency considerations irrelevant once a project is initiated.
4. When the votes are assembled for passage, the sharing of benefits and costs among individuals is no longer a subject open to public review. All projects then serve the national interest.

By contrast, Howe's paper concludes by characterizing ideal decision making, telling us how, in principle, the decisions should be made, and how institutions could be structured so that applied welfare analysis might contribute to the process through appraisal of potential efficiency effects before the decisions, and evaluation of the record afterward. There is little to debate in Howe's ideal framework. But we must remember what Davis's story illustrates—the ideal is not going to work in a representative system of government, and furthermore we should not be surprised by this conclusion.

Public policy does not emerge from a rational planning process. As envisaged in a rational planning setting, the key issue is to convince the mythical policy-maker of the merits of listening to the sage economist. However, the experience of the past three decades and the emergence of the field of public choice in economics call our attention to the importance of accepting what Bromley (1976) referred to as a dialectic view of policy evaluation. Recognizing that there are three stages in the formation of public policy—legislation, rule making, and management—makes it clear that there are many opportunities for individual and coalition interests to be promoted during the process of implementing any public policy. Completion of the process requires that these interests somehow be negotiated and resolved. The overview of environmental legislation of the 1970s by Zeckhauser (1981) describes how this need to resolve conflicting interests influences the features of legislation that is ulti-mately passed. A similar account could likely be given for the complex process of rule making (see Magat, Krupnick, and Harrington [1986]).

The fact that public policymaking is not a rational planning process does not mean there is no role for applied welfare analysis, although it does suggest that we can hope in vain for the rational planning model. Within the actual setting in which policy is formed, the methods of applied welfare analysis provide a framework for defining—in the convenient metric of reductions in net benefits—the consequences of concessions to build consensus, and by so doing they provide tangible measures of the public losses of responding to private interests. Moreover, there remain the prospects for genuine improve-ments in resource allocation decisions, within the constraints on actions at the management stage of the policymaking process.

While it is more difficult to see from the published record, John Krutilla's career reflects a recognition that the greatest prospects for improvement in public policy arise from a mixed strategy. By working at both levels—that is, measuring the losses from proposed and existing programs and at the same time working behind the scenes to help with day-to-day management—im-provements can be made even though public policy formation at times defies a rational planning process. When considered in these terms, Howe's guidelines are not necessarily standards for a rational planning process but methods for developing and disseminating information so that the political consensus un-derlying policy is an informed one.

Progress in understanding the process and in informing its participants requires interaction between two activities: the careful implementation of allocation decisions required by existing statutes as undertaken by the man-agement agency's policy analysts *and* the critical evaluations of the outcomes by analysts outside the controlling agency. Consider the instances discussed by Hyde and Daniels of managerial decisions involving public lands. In one example they provide estimates of the economic losses attributable to the "even-flow" requirement imposed on the Forest Service's timber-harvesting practices. It seems clear that relaxation of these provisions would offer substantial potential for efficiency gains.

The statutory mandates governing agency action in matters of multiple purpose water resource and multiple use land management are complex. It is only an imperfect iterative process by which experts evaluate agency performance and its constraints in relation to the objectives motivating those public actions in the first place. As Cicchetti and Haveman (in this volume) suggest, the institutions that have the responsibility for resolving such differences may not be capable, under their current modes of operation, of dealing with complex economic arguments. As a consequence, there will be costs in this iterative adjustment. Nonetheless, both the Cicchetti–Haveman navigation case and the Hyde–Daniels analysis of the even-flow requirement indicate that the gains are likely to exceed the costs required for appropriate review and evaluation of these critiques.

A second case in the Hyde–Daniels paper illustrates the difficulties of implementing at the managerial level a framework consistent with Howe's proposals. This case involves four campgrounds in the Seeley–Swan Valley in western Montana. For each campground, separable costs must be estimated, joint costs allocated, and distinct demands estimated. Moreover, depending on the nature of these costs, it is entirely possible that a marginal cost pricing policy would imply revenue shortfalls. As the authors readily acknowledge, budget constraints must be taken seriously. Thus, questions concerning the allocation of joint costs not only among different campgrounds, as in this example, but between distinctly different outputs—recreation versus harvested timber—become important in the definition of efficient pricing policies.

Freeman and Norris (in this volume) consider the role of Ramsey pricing rules for such cases. While they cite multiple purpose water projects as the primary area of application for their analysis, it is equally relevant for multiple land use management. Their message is clear: cost allocation and pricing can be simultaneously treated within a framework that views public prices as instruments of policy. Consequently, public prices can be designed to meet objectives in a way that is different from the way prices would be set within a set of idealized market processes.

Ramsey pricing can therefore be treated as a mechanism to assure that a predetermined fraction of costs is recovered efficiently (in a second-best sense) and to allocate joint costs at the same time.[21] Moreover, when either of these objectives is to be met and the activities lead to a public good, then the criteria for efficient production of that good must be altered as well. It must reflect the social cost of Ramsey pricing.

But that is not the end of the story. As observed earlier, the criteria for efficient (in second-best terms) levels of the public good must be enforced. That is, access conditions must assure that the socially optimal level of the public good is in fact provided. This is, of course, the same point that Krutilla

(1966) made in discussing the interaction between efficient allocation and management decisions.

Two further points should be noted. First, Freeman and Norris depart from a substantial portion of the literature on Ramsey pricing by emphasizing the role of the predefined level of cost recovery for the level of the Ramsey prices.[22] Second, in their framework the pricing decision provides the basis for quantifying the social costs of alternative levels of reimbursement by project beneficiaries. They conclude that

> The RBB [Ramsey in my terminology] pricing rules provide a basis for determining the efficient level of cost sharing, that is, the efficient division of total project costs between the purchasers of the project's marketable outputs and taxpayers. The agency should optimally distort its prices upward until it reaches the point where the social cost of raising an additional dollar of revenue through RBB pricing just equals the marginal social cost of raising one dollar of revenue from taxpayers. (Freeman and Norris, p. 127 in this volume)

This conclusion follows because they assume that the extent of difference in a project's costs and operating revenues will be determined in a framework that recognizes the social cost of taxation. As is acknowledged earlier, this is rarely the case. Instead, the political process attempts to distinguish beneficiaries from those paying the costs. Thus, the distortions required by any Ramsey-based reimbursement policies for water projects would more likely be a gauge of the importance of these projects to what Howe describes as the national coalitions required for other programs.

Freeman and Norris also treat the allocation of joint costs as a distinct process within Ramsey pricing. It is important to recognize that the Ramsey approach is a strategy for pricing to recover a specified amount of all costs—separable and joint. It implicitly allocates joint costs but not in the sense usually implied by these processes. Consequently, the Freeman–Norris proposal needs to be seen as one of several possibilities—including the separable-costs–remaining-benefits and the axiomatic approaches—to the allocation of joint costs. The first of these corresponds to conventional practice; it allocates joint costs based on each output's share of the project benefits remaining after the separable costs have been deducted. The second approach specifies a set of axioms that are intended to describe a "fair" cost allocation and then defines the cost assignment implied by those axioms.[23] The axioms generally involve independence from measurement units, cost consistency (i.e., commodities with equal impacts on costs should have the same prices), a requirement of positive prices,[24] and no revenue shortfalls. The pricing schemes that satisfy these five axioms can be treated as a form of average cost pricing.[25]

How do the three approaches compare? It should be said first that there is no single "best" approach. In his overview of the literature on public pricing Bös (1985) points out some differences between two of these methods—

Ramsey and axiomatic prices. They treat cost and demand differently. In the final analysis a choice depends on balancing the importance attached to consumer sovereignty (and our ability to accurately reflect it in conventional demand models) against the principles defining the fair-cost allocations.[26]

We have no experience to date with the performance of reimbursement (i.e., pricing) policies for natural resources outputs. This seems likely to change. The current policy of examining opportunities for privatization of publicly held natural resources has focused renewed attention on the importance of pricing the services of natural and environmental resources provided by public agencies. As a consequence, modifications to existing user fees are under consideration in the planning activities of several resource agencies.

Bowes and Krutilla (in press) suggest that the task of implementing these pricing schemes will need to be modified to account for the special features of the resources involved. For example, in discussing the problem of cost allocation in the context of the multiple use management of public forest lands, they observe:

When a multiple purpose project is put in place in a river development scheme, it will remain in place unchanged unless some modification is introduced by conscious design. This is not the case with the forest management projects where outputs are a function of vegetation manipulation. . . . There are interdependencies among sites as well as between time periods.

The two types of interdependencies combine to pose problems in trying to apply the separable-cost remaining-benefits methods to a given "project." . . . The yield from the site in question cannot be separated from the interacting sites. (Bowes and Krutilla, in press, chap. on "Funding Nonpriced Resource Services")

Their discussion suggests that it may be necessary to consider the design of the size and scope of responsibilities for the operating unit as a part of the pricing decision. That is, the prospects for developing a cost assignment that would be required for efficient pricing itself requires a redefinition of the managerial unit of analysis as well as the outputs delivered by that unit. Thus, insights for the use and the enhancement of applied welfare principles seem to stem from the practice of attempting to implement them in decisions involving natural and environmental resources.

PART II
IMPLICATIONS FOR FUTURE RESEARCH

The influence of resource and environmental problems on applied welfare economics is not likely to end in the near future. Many of the problems facing current policymakers stretch the limits of our understanding of the tenets of welfare economics as well as the realization of those principles in the operating and investment rules for managing and allocating natural and environmental resources. Three of the current challenges in this field seem especially relevant to the work discussed in this volume and are therefore singled out for further discussion. They are (1) axiomatic approaches to applied welfare economics, (2) policy control of uncertainty, and (3) natural resource damage assessments.

Axiomatic Approaches to Applied Welfare Economics

Benefit–cost analysis was originally viewed as a set of procedures for evaluating public investment projects that primarily affected people living in the same generation as the policymakers who made the decisions. Today, in contrast, we face decisons about problems with exceptionally long-term consequences— some extending beyond the time span of any known civilization. Disposal of commercial nuclear waste, environmental problems resulting from pollutants such as carbon dioxide or chlorofluorocarbons, and the consequences of exploitation of nonrenewable resources are in varying degrees all examples of this type of problem. Can the existing method of benefit–cost analysis be adapted to deal with such problems, or do we need a completely different perspective?

There seem to be at least three positions on this question emerging from the literature. From the first comes the title of this subsection in my description of future research. It suggests a reorientation in our analysis of social welfare functions—a shift from emphasis on the impossibility of realizing prespecified objectives as part of a collective decision process to a focus on alternative sets of axioms that lead to operational rules for such long-term decisions. Such a framework is comparable to the axiomatic approach to cost allocation. Clearly, it is the logical next step to follow from the arguments of Page (in this volume) concerning intergenerational equity. It is unlikely that there would be broad agreement on any set of axioms.[27] Nevertheless, evaluating the implications of the decision criteria implied by such axiom systems as compared with the implications of more conventional benefit–cost analysis of such policy actions would help determine whether such consensus had policy significance. Page's analysis indicates that it can have significance for decisions on problems involving infinite time horizons.

A second proposed approach to these problems is an adaptation of some of the findings of Krutilla's program of research associated with unique natural environments. According to this view, the policy environment of all society is short term in relation to the extended time horizon of such problems. This is especially true for the United States. Therefore, given the uncertainties that surround decisions on these matters, the criteria used in evaluating them should recognize the special role of new information as well as the effect current decisions might have on the value of pertinent information that comes to light in the future. This alternative would simply adapt the general approach described by Fisher (in this volume), together with other related work (see Arrow and Fisher [1974]; Fisher and Hanemann [1986]; Hanemann [1984]) to deal with long-term decisions. Its value to practical decision making hinges on the ability of analysts to determine what the conditional value of new information might be under alternative decisions and developments in the world.

Finally, the last approach considers such decisions outside of economics, based instead on the ethical principles that a society wishes its policymakers to adopt with respect to future generations. The difficulty here is that there is no one generally accepted ethical system, but many. As Kneese and Schulze (1985) observe in reviewing the role of the literature on environmental ethics for the theory and practice of applied welfare economics, political decision processes resolve value conflicts; they are not organized in response to a single value system.

Regardless of the criteria used by the policymaker in arriving at them, all decisions implicitly assign values to the outputs provided. Consequently, the literature on environmental ethics is misleading to suggest that we can avoid assigning values to the outputs (or to the outcomes) of a decision process by simply recommending ethical structures for policymakers. At best such thinking provides some justification for constraints on what this generation can do, the purpose of which are to maintain flexibility in the future (see Ciriacy-Wantrup [1952] and Page [1977a] for an alternative rationale for such constraints).

It is clear that in these cases, *ex post* evaluations of alternative decisions are impossible. Furthermore, the temporal structure of these problems precludes actual experimentation. But that fact does not, in itself, prevent analysis of the implications of alternative decision rules. Indeed, it should enhance the value of such efforts. That is, comparisons that help explain how the inherent structure of such long-term problems affects the decisions recommended by alternative evaluation methods should have considerable value regardless of the ethical system to which one adheres. Such information serves to inform policy debates.

To date, most of the work in this area has been at a conceptual level, with much less research focused on detailed analyses of specific policy actions. As with the specific evaluations of water resource projects in the late fifties and early sixties, such applications would probably stimulate advances in theory and in new practical methods for dealing with these types of problems.

Policy Control of Uncertainty

Evaluations of flood control projects were among the first to explicitly recognize the possibility of valuing policies that change the risks facing individuals.[28]

These efforts generally defined the benefits of such projects as the expected value of the losses that individuals avoided as a result of a control program's change in the probabilities of alternative levels of damages (see Herfindahl and Kneese [1974] for a summary). Current literature on the valuation of changes in risk would question this choice of a benefit concept, but it should be acknowledged that this early literature did identify both its proposed valuation concept and its counting on the ability of lay people to accurately perceive probabilities as important assumptions deserving further study (see, for example, Burton, Kates, and White [1978, especially chap. 4]). As a consequence, the flood control work provides a clear foreshadowing of the current research issues identified by Kneese (in this volume). How should the valuation concepts and decision criteria be adapted to deal with policies that deliver reductions in risks with long latency periods, such as those associated with hazardous wastes? In many respects these policy decisions involve most of the troublesome aspects identified with the case of flood control policies, namely, changes in very small risks, where the outcomes at risk have potentially disastrous consequences for the individual.

The research issues raised by such problems bridge the concepts associated with nonuse values (identified with Krutilla's early work on unique natural environments) and questions of the appropriate public attitude toward uncertainty. As is now recognized, an expected utility framework would imply that the relevant monetary concept of the benefits provided by a project reducing the risk of a detrimental event is that value which *holds expected utility constant*, given the opportunities available for individual adjustment. With the assumption of state-dependent preferences, this value will not necessarily correspond to the expected monetary value of damages avoided (see Cook and Graham [1977]).

Implementation of this alternative valuation framework is further confounded by evidence from cognitive psychology indicating that individuals act on incomplete and erroneous information, rely on suboptimal methods for assessing risk (often referred to as heuristics), and do not understand the limits of their own knowledge (see Slovic, Fischhoff, and Lichtenstein [1985] for an overview).

It also appears that individuals' behavior is affected by both the attributes of the risk (e.g., its long-latency or voluntary nature) and of the events at risk. While some encouraging evidence indicates that the learning process from risk information is rational, it seems clear that our conceptualization of risk perception and valuation will need to be integrated before a complete explanation of existing behavior will be possible (see Smith and Johnson [1988]; Smith and coauthors [1987]; and Viscusi and O'Connor [1984] for more details). In several areas of environmental policy there is additional motivation for this line of research. Information programs may well be the only feasible policy instruments for dealing with potentially important environmental sources of health risks.

Relying on these programs to meet policy objectives of protecting human health implies that we understand how to explain these risks and, furthermore, have confidence that, once informed, individuals will make sound judgments.

At present, the empirical evidence supporting consistent behavior is limited. Nonetheless, it does appear that we should be able to understand how individuals respond to risk information (see Smith and coauthors [1987]). This implies that it should be feasible to test a framework incorporating risk perception with a description of behavioral choice under uncertainty. Within such a framework, valuation concepts and measures of risk aversion represent alternative methods for describing how an individual responds to changes in risk. The problem does not end here, however; the type of risk and conception of society's obligations will determine how individuals' valuations of risk changes are to be used in informing society's decisions (see Ulph [1982]).

With the growing importance of risk management in environmental and other forms of social regulation, developing a consistent and empirically viable framework for valuation can be expected to occupy a large portion of the attention of future research in applied welfare economics.

Natural Resource Damage Assessments

The Comprehensive Environmental Response, Compensation and Liability Act (CERCLA) of 1980 and its reauthorization—the Superfund Amendments and Reauthorization Act (SARA) of 1986—require the promulgation of regulations to assess the damages involving the release of hazardous wastes into natural resources. The statute defines natural resources broadly to include land, fish, wildlife, biota, air, water, groundwater, drinking-water supplies, and natural or environmental resources under the jurisdiction of a government trustee.

Several important research questions in this area are now developing. The most important of them arises from the requirement of estimating the damages from an injury using a defined baseline and of estimating a restoration or replacement cost for the natural resource. The rules provide fairly broad guidelines for determining the baseline: it is to represent the conditions that were present prior to the injury, regardless of the historical point when the injury occurred. For some current cases involving mines in Colorado and the contamination of rivers adjoining the mines, this approach implies that damages must be estimated from the late 1800s or early 1900s.

The problem posed by such calculations is somewhat similar to those raised by projects with resource commitments extending over many generations. Issues of consistency in preferences, capitalization of damages to current-period dollars, and the selection of a rate of return for that capitalization are a few of the most important considerations. There is, of course, an important difference. Historical records of events of these times and indirect evidence on the nature of households' preferences do exist.

Although CERCLA and SARA do not consider whose preferences are to be used, the analysis must. The legislation reflects environmental concerns that were not universally held thirty years ago. Knowledge of the consequences of such contamination as well as attitudes toward the use of natural resources were quite different then. If the objective is to provide an equivalent asset or set of services, what group is to define the nature of the equivalence in such long-term retrospective appraisals?

Equally important, the objective is usually to evaluate the diminution in the use values (broadly defined)[29] available to each type of user (excluding commercial uses) from the time of the incident leading to the damage. The damage is to be evaluated using replacements costs (for the services lost) or damage assessment methods (figuring losses in consumer benefits), whichever is less. Restoration or replacement costs are also to be used in evaluating the current status of the resource. To the extent that the present value of user benefits is less than either of these costs, it is to be used in evaluating the injury to the resource.

Estimation of the change in the value of a natural resource as an asset also raises interesting issues: the selection of a discount rate and the desirability of symmetrical treatment of the rate used in appreciating past losses versus that for future damages are among these.

Once again, the issues described in the damage assessment practices relate directly to concepts developed in past research applying welfare principles to allocation decisions involving natural resources. Notable among these concepts are Krutilla's contributions.

More examples of future research avenues could be cited with similar links to the stimulus provided by real policy issues, but the discussion as it stands should suffice to make the point: the interaction between theory and practice has *definitely* been a two-way street. Moreover, most of the important conceptual issues in applied welfare economics as it has been practiced in the United States over the past three decades can be traced to resource and environmental problems. John Krutilla's work has touched nearly all aspects of this evolution—a truly remarkable record.

Acknowledgment

Thanks are due Clive Bell for helpful discussions of the interaction between the theory and application of benefit–cost analyses as they developed in the United Kingdom, and to Clifford Russell, William Desvousges, anonymous reviewers, and the authors of the other papers in this volume, especially Robert Haveman, for most helpful comments on earlier drafts of this paper. Partial support for this research was provided by the Centennial Professor Fund of Vanderbilt University.

NOTES

1. More specifically, he observed:

I do not mean to imply that we must throw up our hands altogether and take no action of any sort on the practical questions concerned because our analytic structure is in too highly imperfect a state to enable us to give categorical answers in our role as economic theorists. Such a council of paralysis through despair is not my intent. I believe that the politician is, in many cases, justified in taking, and indeed forced to take action on many such questions, perfect analysis or no. . . . Nor do I suggest that the politician cannot receive useful assistance in these matters from the judgement and observation of the "practical" economist. (Baumol, 1967, p. 207)

As Krutilla (1981) documents, Baumol was not alone in these judgments. Scitovsky (1954) expressed comparable doubts about the prospects for applying externality theory to real-world problems.

2. The presidential address by Krutilla (1981) does not go this far; it simply indicates that resource and environmental problems have provided particularly "effective vehicles" for the creative application of what has been learned in welfare economics.

3. This statement seems most relevant to benefit–cost analysis developed in the policy environment of the United States. Economists working on the theory of benefit–cost analysis in the United Kingdom seem to have been more directly influenced by applications in the context of developing countries. See the recent reviews by Dreze and Stern (forthcoming) and Squire (1986) for overviews of this literature. Certainly the focus in non-U.S. studies on the theory of shadow prices and the importance of distortions to project evaluation is a reflection of their importance for public investment decisions in developing economies.

4. I have implicitly adopted a household production framework and thus consider this description to include interventions in household activities as well.

5. The Kaldor–Hicks potential compensation test simply requires that the aggregate of benefits received exceed the aggregate costs. In this case beneficiaries could, in principle, compensate those experiencing losses and still realize gains. Most discussions of this criterion generally include consideration of the prospect of Scitovsky–Samuelson reversals. See Just, Hueth, and Schmitz (1982, chap. 3) for further elaboration.

6. The current requirement of President Reagan's Executive Order 12291 of 1980 calls for benefit–cost analyses to be conducted for any new regulations or revisions to existing regulations that are major, with *major* defined as imposing a $100-million annual cost on the economy. This would seem, even for an economy as large as that of the United States, to call for applying the

method outside the small-scale cases it was originally intended to address. See Smith (1984b) for further discussion.

7. Randall (1986) has recently offered a straightforward description of the conceptual rationale for benefit–cost analysis, observing:

BCA [benefit–cost analysis] is a test for PPI's [potential Pareto improvements]. The pragmatic task in BCA is to employ appropriate empirical methods within an economic-theoretic framework consistent with the PPI concept, to determine ex ante whether it is likely that a proposed change would, if implemented, generate a PPI. (p. 169)

8. The use of *authentic* rather than *accurate* is deliberate. At issue is whether it is possible to approximate the features of these processes that are important to the decisions involved. Thus, the terminology implicitly assumes that we will never know the "true" technology or preference structure, so what is important is whether we make large systematic errors in our descriptions of them.

9. The Fisher–Krutilla–Cicchetti (1972) model viewed society's objective as one of maximizing the present value of the economic surplus generated by two potential allocations of a specified natural environment. One of these involved a developed use of the resource while the second left the resource in its original state.

Their original application dealt with a hydroelectric facility on the Hells Canyon reach of the Snake River. In a preserved state, this area provided recreational and amenity values. For discussion of other cases within this general framework, see Krutilla and Fisher (1985).

10. This dictum is similar to Stiglitz's questioning the role for policy in the case of nonrenewable resources by observing that

the existence of a natural resource problem [i.e., an impending exhaustion of essential natural resources] has no immediate implications: it is neither a necessary nor a sufficient condition for governmental intervention in the markets for natural resources. The market could be doing as well as could be done and the economy could still be facing a doomsday. . . . (Stiglitz, 1979, p. 49)

11. As observed earlier, the same type of bidirectional exchange appears to characterize the relationship between the developments in the methodology of benefit–cost analysis and the problems of public investment in developing economies for research in the United Kingdom on valuation of nonmarketed resources.

12. This point is valid even if we assume that an individual can have either use or existence values but not both. The nature of the activities undertaken to derive utility from the resource will be different in the two cases.

13. See Cook and Graham (1977) for the definition of an index of uniqueness and Smith (1984a) and Plummer and Hartman (1986) for use of this concept in describing the magnitude of option value.

14. This is consistent with the description by Zeckhauser (1981) of the process of developing legislative consensus. For discussion of its implications, see Smith (1984b).

15. This has been a consistent theme in Krutilla's work on both investment and management decisions for public resources. Implementation of an efficiency mandate requires that all factors influencing the values of the services provided by the resources under study be considered in the evaluation of investment and management policies. These factors would include (1) the interactions between the conditions of access to public resources, the implied levels of use, and the resulting quality of services; (2) the connections in production between decisions on the multiple private and public outputs provided by a specific resource; and (3) the implications of any institutional restrictions designed to accommodate nonefficiency goals on allocation or management decisions. The paper by Hyde and Daniels (in this volume) clearly illustrates the importance of the third of these considerations. Indeed, these authors' first example suggests that restrictions on the harvesting decisions on public forest lands (i.e., the "even-flow" mandates) force public agencies to ignore market signals and result in rather dramatic opportunity costs—in terms of both consumer surpluses and government revenues.

16. The simple repackaging hypothesis refers to the assumption that individuals' attitudes toward the characteristics of goods they consume is the simple aggregate of each characteristic that affects utility. This would imply that two eight-room homes and one sixteen-room home would yield equivalent utility. This is clearly implausible in some situations; in others it may offer a reasonable description of individuals' decisions on hetero-geneous commodities.

17. For a comprehensive overview of this literature to about 1980, see Lind (1982).

18. This contrasts with the European experience where public ownership (as opposed to regulation) is more common in a range of production activities that produce private goods. The introduction to the paper by Bös (1985) on public pricing describes the rationale for such intervention and provides a brief overview of the sectors involved in selected European countries.

19. This distinction parallels the dichotomy between cost allocation and the design of a pricing structure that Cicchetti and Haveman (in this volume) describe in discussing their electricity rate design case.

20. This is certainly consistent with the conclusion presented by Young and Haveman (1985) in their overview of the economic issues associated with water resource projects. Indeed, they observe that

Improvements in the efficiency and equity of water resources policy are likely to be achievable by a reduction of the large volume of subsidies conferred by existing policy. Elimination of the existing separation of beneficiaries and cost-bearers of policy measures through a comprehensive beneficiary charge policy could yield this improvement. (p. 521)

21. The recent book by Baumol (1986) on superfairness concepts takes strong exception to this interpretation, observing instead:

The point is that an allocation of costs is a process for subdivision of the firm's total cost that, unlike the Ramsey solution, aspires to find some a priori formula, based *only* on costs and output quantities, that assigns to each product its proper share of total cost. It is then only by distortion of language that every pricing procedure that brings in enough revenue overall to cover total cost is defined, from that attribute alone, to constitute a cost allocation. (p. 144)

The relevance of Baumol's argument to the cases considered by Freeman and Norris seems limited because a cost allocation is simply an intermediate step in deriving beneficiary charges or prices. Moreover, as Baumol acknowledged earlier, even in situations where marginal cost pricing can be applied, only when all marginal costs (i.e., those for all outputs) are independent of the output vector will the cost allocation be independent of demand. That is, we must know the output vector implied by each set of prices to calculate marginal costs, if the prices are to be compatible with a market equilibrium. (See Baumol [1986], pp. 140–142.)

22. Bös (1985) discusses the case of Ramsey pricing with interdependent demands for private goods.

23. See Aumann and Shapley (1974) or Young (1985) for discussion.

24. This can be interpreted as implying they should be positive for any product that requires investment.

25. Baumol's (1986) position on all axiomatic cost allocation methods is that they are arithmetic processes associated with cost accounting and cannot bear a relationship to fairness because they leave out consideration of consumer demands.

26. Bös (1985) identifies three issues in relating Ramsey and axiomatic cost allocation (and pricing) methods. Ramsey prices consider marginal costs at the "optimal" outputs; axiomatic methods consider marginal costs at all outputs. Ramsey methods are compatible with a market equilibrium; axiomatic prices are not necessarily compatible. Ramsey prices often require a detailed and plausible characterization of consumers' demands in the relevant outputs; axiomatic methods do not.

27. The criticism by Baumol (1986, pp. 147–149) of Loehman-Whinston (1974) is an example of such disagreements.

28. For an overview, see Herfindahl and Kneese (1974).

29. See Kopp and Smith (1988) for a detailed discussion of the legislation, Department of the Interior rules for natural resource damage assessments, and selected examples to illustrate how these factors are related in practice.

REFERENCES

Arrow, Kenneth J. 1963. *Social Choice and Individual Values*. 2d ed. (New York, Wiley).

————, and Anthony C. Fisher. 1974. "Environmental Preservation, Uncertainty, and Irreversibility," *Quarterly Journal of Economics* vol. 88 (May) pp. 312–319.

————, and Robert Lind. 1970. "Uncertainty and the Evaluation of Public Investment Decisions," *American Economic Review* vol. 60, pp. 364–378.

Aumann, R. J., and L. S. Shapley. 1974. *Values of Non-Atomic Games* (Princeton, N.J., Princeton University Press).

Baumol, William J. 1967. *Welfare Economics and the Theory of the State*. 2d ed. (Cambridge, Mass., Harvard University Press).

————. 1986. *Superfairness: Applications and Theory* (Cambridge, Mass., MIT Press).

Bohm, Peter, and Clifford S. Russell. 1985. "Comparative Analysis of Alternative Policy Instruments," in A. V. Kneese and J. L. Sweeney, eds., *Handbook of Natural Resource and Energy Economics* vol. I (Amsterdam, North-Holland).

Bös, Dieter. 1985. "Public Sector Pricing," in A. J. Auerbach and M. Feldstein, eds., *Handbook of Public Economics* (Amsterdam, North-Holland).

Bowes, Michael D., and John V. Krutilla. 1985. "Multiple Use Management of Public Forestlands," in A. V. Kneese and J. L. Sweeney, eds., *Handbook of Natural Resource and Energy Economics* vol. II (Amsterdam, North-Holland).

————, and ————. In press. *Multiple-Use Management: The Economics of Public Forestlands* (Washington, D.C., Resources for the Future).

Boyle, K. J., and R. C. Bishop. 1987. "The Total Value of Wildlife: A Case Study Involving Endangered Species," *Water Resources Research* vol. 23 (May) pp. 943–950.

Bradford, David. 1970. "Benefit–Cost Analysis and Demand Curves for Public Goods," *Kyklos* vol. 23, pp. 1145–1159.

Bromley, Daniel W. 1976. "Economics and Public Decisions: Roles of the State and Issues in Economic Evaluation," *Journal of Economic Issues* vol. 10 (December) pp. 811–838.

Brookshire, David S., Mark A. Thayer, William D. Schulze, and Ralph C. d'Arge. 1982. "Valuing Public Goods: A Comparison of Survey and Hedonic Approaches," *American Economic Review* vol. 72 (March) pp. 165–177.

Burton, Ian, Robert W. Kates, and Gilbert F. White. 1978. *The Environment as Hazard* (New York, Oxford University Press).

Cornes, Richard, and Todd Sandler. 1986. *The Theory of Externalities, Public Goods, and Club Goods* (Cambridge, Cambridge University Press).

Cicchetti, Charles J., and V. Kerry Smith. 1976. *The Costs of Congestion* (Cambridge, Mass., Ballinger).

Ciriacy-Wantrup, S. V. 1952. *Resource Conservation* (Berkeley, University of California Press).

Cook, Phillip J., and Daniel A. Graham. 1977. "The Demand for Insurance and Protection: The Case of Irreplaceable Commodities," *Quarterly Journal of Economics* vol. 91 (February) pp. 143–156.

Cropper, Maureen L. 1976. "Regulating Activities with Catastrophic Environmental Effects," *Journal of Environmental Economics and Management* vol. 3 (June) pp. 1–15.

Cummings, Ronald G., David S. Brookshire, and William D. Schulze. 1986. *Valuing Public Goods: The Contingent Valuation Method* (Totowa, N.J., Rowman and Allenheld).

d'Arge, Ralph C., and K. C. Kogiku. 1973. "Economic Growth and the Environment," *Review of Economic Studies* vol. 40 (January) pp. 61–78.

Dreze, Jean, and Nicholas Stern. Forthcoming. "The Theory of Cost–Benefit Analysis," in A. J. Auerbach and M. Feldstein, eds., *Handbook of Public Economics* vol. II (Amsterdam, North-Holland).

Edelson, N. 1971. "Congestion Tolls Under Monopoly," *American Economic Review* vol. 61 (September) pp. 871–882.

Ferejohn, John, and Talbot Page. 1978. "On the Foundations of Inter-Temporal Choice," *American Journal of Agricultural Economics* (May) pp. 15–21.

Fisher, Anthony C. 1973. "Environmental Externalities and the Arrow–Lind Public Investment Theorem," *American Economic Review* vol. 63 (September) pp. 722–725.

————. 1981. *Resource and Environmental Economics* (Cambridge, Cambridge University Press.

————, and W. Michael Hanemann. 1986. "Option Value and the Extinction of Species," in V. K. Smith, ed., *Advances in Applied Microeconomics* vol. 4 (Greenwich, Conn., JAI Press).

————, and John V. Krutilla. 1972. "Determination of Optimal Capacity of Resource Based Recreation Facilities," *Natural Resources Journal* vol. 12 (July) pp. 417–444.

————, and John V. Krutilla. 1975. "Conservation, Environment, and the Rate of Discount," *Quarterly Journal of Economics* vol. 89 (August) pp. 358–370.

————, and John V. Krutilla. 1985. "Economics of Nature Preservation," in A. V. Kneese and J. L. Sweeney, eds., *Handbook of Natural Resource and Energy Economics* vol. I (Amsterdam, North-Holland).

————, John V. Krutilla, and Charles J. Cicchetti. 1972. "The Economics of Environmental Preservation: A Theoretical and Empirical Analysis," *American Economic Review* vol. 62 (September) pp. 605–619.

Freeman, A. Myrick, III. 1979. *The Benefits of Environmental Improvement: Theory and Practice* (Baltimore, Md., Johns Hopkins University Press for Resources for the Future).

————. 1985. "Methods for Assessing the Benefits of Environmental Programs," in A. V. Kneese and J. L. Sweeney, eds., *Handbook of Natural Resource and Energy Economics* vol. I (Amsterdam, North-Holland).

————, and Robert H. Haveman. 1977. "Congestion, Quality Deterioration and Heterogeneous Tastes," *Journal of Public Economics* vol. 8, pp. 225–232.

Hanemann, W. Michael. 1984. "Information and the Concept of Option Value." Working Paper. University of California, Berkeley, Department of Agricultural and Resource Economics.

Haveman, Robert H. 1970. "Public Expenditures and Policy Analysis: An Overview," in R. H. Haveman and J. Margolis, eds., *Public Expenditures and Policy Analysis* (Chicago, Ill., Markham Publishing).

————, and John V. Krutilla, with Robert M. Steinberg. 1968. *Unemployment, Idle Capacity, and the Evaluation of Public Expenditures: National and Regional Analyses* (Baltimore, Md., Johns Hopkins Press for Resources for the Future).

Herfindahl, Orris C., and Allen V. Kneese. 1974. *Economic Theory of Natural Resources* (Columbus, Ohio, Charles E. Merrill).

Johnson, M. B. 1962. "On the Economics of Road Congestion," *Econometrica* vol. 32, pp. 137–150.

Just, Richard E., Darrell L. Hueth, and Andrew Schmitz. 1982. *Applied Welfare Economics and Public Policy* (Englewood Cliffs, N.J., Prentice-Hall).

Karni, Edi. 1983. "Risk Aversion for State-Dependent Utility Functions: Measurement and Applications," *International Economic Review* vol. 24 (October) pp. 637–648.

————. 1987. "Generalized Expected Utility Analysis of Risk Aversion with State Dependent Preferences," *International Economic Review* vol. 28 (February) pp. 229–240.

Kneese, Allen V., and William D. Schulze. 1985. "Ethics and Environmental Economics," in A. V. Kneese and J. L. Sweeney, eds., *Handbook of Natural Resource and Energy Economics* vol. I (Amsterdam, North-Holland).

Koopmans, Tjalling. 1972. "Two Papers on the Representation of Preference Ordering with Independent Components of Consumption and Representation of Preference Orderings Over Time." Cowles Foundation Paper No. 366.

Kopp, Raymond J., and V. Kerry Smith. 1988. *Can Natural Resource Damage Assessments Be Performed? A Summary of Economics Issues.* Discussion Paper No. QE88-03 (Washington, D.C., Resources for the Future).

Krutilla, John V. 1966. "Is Public Intervention in Water Resources Development Conducive to Economic Efficiency?" *Natural Resources Journal* vol. 6 (January) pp. 60–75.

————. 1967. "Conservation Reconsidered," *American Economic Review* vol. 57 (September) pp. 777–786 (Resources for the Future Reprint No. 67).

————. 1969. "Efficiency Goals, Market Failure, and the Substitution of Public for Private Action," in U.S. Congress, Joint Economic Committee, *The Analysis and Evaluation of Public Expenditures: The PPB System* vol. I (Washington, D.C., Government Printing Office).

————. 1981. "Reflections of an Applied Welfare Economist," *Journal of Environmental Economics and Management* vol. 8 (March) pp. 1–10.

————, and Otto Eckstein. 1958. *Multiple Purpose River Development: Studies in Applied Economic Analysis* (Baltimore, Md., Johns Hopkins Press for Resources for the Future).

————, and Anthony C. Fisher. 1985. *The Economics of Natural Environments: Studies in the Valuation of Commodity and Amenity Resources.* Rev. ed. (Washington, D.C., Resources for the Future).

Lind, Robert C. 1982. "A Primer on the Major Issues Relating to the Discount Rate for Evaluating National Energy Options" in Robert C. Lind and co-authors, *Discounting for Time and Risk in Energy Policy* (Washington, D.C., Resources for the Future).

Loehman, Edna, and Andrew Whinston. 1974. "An Axiomatic Approach to Cost Allocations for Public Investment," *Public Finance Quarterly* (April) pp. 236–251.

Lucas, Robert C. 1964. "Wilderness Perception and Use: The Example of the Boundary Waters Canoe Area," *Natural Resources Journal* vol. 3, pp. 394–411.

Maass, Arthur. 1966. "Benefit–Cost Analysis: Its Relevance to Public Investment Decisions," *Quarterly Journal of Economics* vol. 80 (May) pp. 208–226.

Magat, Wesley A., Alan J. Krupnick, and Winston Harrington. 1986. *Rules in the Making* (Washington, D.C., Resources for the Future).

Mäler, Karl-Göran. 1974. *Environmental Economics: A Theoretical Inquiry* (Baltimore, Md., Johns Hopkins University Press for Resources for the Future).

McConnell, Kenneth E. 1983. "Existence and Bequest Value," in R. D. Rowe and L. G. Chestnut, eds., *Managing Air Quality and Scenic Resources at National Parks and Wilderness Areas* (Boulder, Colo., Westview Press).

Mitchell, Robert Cameron, and Richard T. Carson. 1988. *Using Surveys to Value Public Goods: The Contingent Valuation Method* (Washington, D.C., Resources for the Future).

Page, Talbot. 1973. *Economics of Involuntary Transfers* (Berlin, Springer Verlag).

———. 1977a. *Conservation and Economic Efficiency: An Approach to Materials Policy* (Baltimore, Md., Johns Hopkins University Press for Resources for the Future).

———. 1977b. "Equitable Use of the Resource Base," *Environment and Planning*, vol. 9, pp. 15–22.

Plummer, Mark L., and Richard C. Hartman. 1986. "Option Value: A General Approach," *Economic Inquiry* vol. 24 (July) pp. 455–471.

Porter, Richard C. 1982. "The New Approach to Wilderness Preservation Through Benefit–Cost Analysis," *Journal of Environmental Economics and Management* vol. 9 (March) pp. 59–80.

Randall, Alan. 1985. "The Total Value Dilemma." Unpublished paper prepared for U.S. Forest Service Workshop on Non-Use and Intrinsic Values. Revised June 30, 1985.

———. 1986. "Valuation in a Policy Context," in D. A. Bromley, ed., *Natural Resource Economics: Policy Problems and Contemporary Analysis* (Boston, Mass., Kluwer-Nijhoff).

———, and John R. Stoll. 1983. "Existence Value in a Total Valuation Framework," in R. D. Rowe and L. G. Chestnut, eds., *Managing Air Quality and Scenic Resources at National Parks and Wilderness Areas* (Boulder, Colo., Westview Press).

Scitovsky, Tibor. 1954. "Two Concepts of External Economics," *Journal of Political Economy* vol. 62 (2) pp. 143–151.

Shechter, Mordechai, and Robert C. Lucas. 1978. *Simulation of Recreational Use for Park and Wilderness Management* (Baltimore, Md., Johns Hopkins University Press for Resources for the Future).

Slovic, Paul, Baruch Fischhoff, and Sarah Lichtenstein. 1985. "Regulation of Risk: A Psychological Perspective," in Roger G. Noll, ed., *Regulatory Policy and the Social Sciences* (Berkeley, University of California Press).

Smith, V. Kerry. 1984a. "A Bound for Option Value," *Land Economics* vol. 60 (August) pp. 292–296.

————, ed. 1984b. *Environmental Policy Under Reagan's Executive Order* (Chapel Hill, N.C., University of North Carolina Press).

————. 1985. "Intrinsic Values in Benefit–Cost Analysis." Unpublished paper prepared for U.S. Forest Service Workshop on Non-Use and Intrinsic Values, Vanderbilt University, Nashville, Tenn. Revised March 1985.

————. 1987. "Uncertainty, Benefit–Cost Analysis and the Treatment of Option Value," *Journal of Environmental Economics and Management* (September).

————, and F. Reed Johnson. 1988. "How Do Risk Perceptions Respond to Information? The Case of Radon," *Review of Economics and Statistics* vol. 70 (February) pp. 1–8.

————, and John V. Krutilla. 1976. *Structure and Properties of a Wilderness Travel Simulator: An Application to the Spanish Peaks Area* (Baltimore, Md., Johns Hopkins University Press for Resources for the Future).

————, and John V. Krutilla. 1979. "Resource and Environmental Constraints to Growth," *American Journal of Agricultural Economics* vol. 61 (August) pp. 395–408.

————, William H. Desvousges, Ann Fisher, and F. Reed Johnson. 1987. *Communicating Radon Risk Effectively: A Mid-Course Evaluation*. Report to U.S. Environmental Protection Agency, Vanderbilt University, Nashville, Tenn., July.

Spence, Michael. 1984. "Cost Reduction, Competition, and Industry Performance," *Econometrica* vol. 52 (January) pp. 101–122.

Squire, Lyn. 1986. "Project Evaluation in Theory and Practice," Unpublished paper, The World Bank, Washington, D.C., January.

Stankey, George H. 1971. "The Perception of Wilderness Recreation Carrying Capacity: A Geographic Study in Natural Resources Management" (Ph.D. dissertation, Michigan State University).

Stiglitz, Joseph E. 1979. "A Neoclassical Analysis of the Economics of Natural Resources," in V. K. Smith, ed., *Scarcity and Growth Reconsidered* (Baltimore, Md., Johns Hopkins University Press for Resources for the Future).

Tietenberg, Thomas H. 1986. *Emissions Trading: An Exercise in Reforming Pollution Policy* (Washington, D.C., Resources for the Future).

Ulph, Alistair. 1982. "The Role of Ex Ante and Ex Post Decisions in the Valuation of Life," *Journal of Public Economics* vol. 18, pp. 265–276.

Viscusi, W. Kip, and Charles J. O'Connor. 1984. "Adaptive Responses to Chemical Labeling: Are Workers Bayesian Decision Makers?" *American Economic Review* vol. 74 (December) pp. 942–956.

Walters, Alan A. 1961. "The Theory and Measurement of Private and Social Costs of Highway Congestion," *Econometrica* vol. 29, pp. 676–699.

Weisbrod, Burton A. 1964. "Collective Consumption Services of Individual Consumption Goods," *Quarterly Journal of Economics* vol. 78 (August) pp. 471–477.

Wilman, Elizabeth A. 1986. "Valuing Quality Changes in Recreational Resources." Unpublished paper, University of Calgary, Canada.

Young, H. Peyton. 1985. "Methods and Principles of Cost Allocation," in H. P. Young, ed., *Cost Allocation: Methods, Principles, Applications* (Amsterdam, Elsevier Science).

Young, Robert A., and Robert H. Haveman. 1985. "Economics of Water Resources: A Survey," in A. V. Kneese and J. L. Sweeney, eds., *Handbook of Natural Resource and Energy Economics* vol. II (Amsterdam, North-Holland).

Zeckhauser, Richard. 1981. "Preferred Policies When There Is a Concern for Probability of Adoption," *Journal of Environmental Economics and Management* vol. 8 (September) pp. 215–237.

THREE DECADES OF
WATER RESOURCES RESEARCH:
A PERSONAL PERSPECTIVE

Allen V. Kneese

John Krutilla and I have been friends and associates at Resources for the Future much longer than either of us cares to remember. But John's influence on me and on my professional development started even before I knew him personally. The first economics book that I read after graduate school, other than textbooks for the courses I was teaching, was the volume that he and Otto Eckstein published in 1958: *Multiple Purpose River Development: Studies in Applied Economic Analysis*. It was only the second book about water resources to come out of an exciting new organization called Resources for the Future (RFF). At that time I was a fledgling assistant professor of economics at the University of New Mexico, and, among the many other pieces of luck that have been my good fortune, my department chairman there was Nathaniel Wollman, then still in the early years of his career, later one of the nation's distinguished figures in water resources economics. He knew John, and lent me a copy of *Multiple Purpose River Development*.

During that time Nat Wollman and I worked on a project funded by the first grant to an outside body made by RFF in the water resources area. The study dealt with the economics of allocating New Mexico's share of the Upper Basin entitlement to water from the Colorado River under the terms of the Law of the River. To give younger readers an idea of how long ago this was, if not in elapsed time at least in terms of technology, one of the capital items we acquired for this project was a brand new Marchant mechanical desktop calculator. We were very proud of it because it had a motor and therefore

Author's note: This essay is a personal perspective in the sense that its objective is not comprehensive coverage of the field. My citations to the literature reflect this perspective, which is focused mostly on the works of John V. Krutilla and Resources for the Future, and therefore do not include many other valuable contributions.

one did not have to pull a handle to set the gears whirring. This study resulted (after a long publication delay) in the book *The Value of Water in Alternative Uses* (Wollman, 1962).

The Project Evaluation Era

The late 1950s and early 1960s saw the publication of a number of classics in water resources economics. The first of these represented both the peak and the end of an era—that of studies of the application of benefit–cost analysis to individual water projects. The book was *Water Resources Development: The Economics of Project Evaluation* by Otto Eckstein (1958). This study, which is worth reading even today, was the academic capstone of an era in a field of applied economics that was developed, not by the university or research community, but by economists and engineers working in federal agencies responsible for the economic evaluation of public projects.

In 1808, Jefferson's Secretary of the Treasury, Albert Gallatin, brought out his report on a transportation (navigation) program for the new nation, and from that time to the present, public water development agencies have found it necessary and desirable to systematically compare estimated benefits with the costs of proposed development projects. The federal Reclamation Act of 1902 required economic analysis of projects; the Flood Control Act of 1936 established the welfare economics feasibility test that benefits "to whomsoever they may accrue" must exceed costs. In 1946 the Federal Interagency River Basin Committee appointed a Subcommittee on Benefits and Costs to reconcile the practices of federal agencies in making benefit–cost analyses. Four years later this subcommittee issued a landmark report entitled *Proposed Practices for Economic Analysis of River Basin Projects* (Federal Interagency River Basin Committee, Subcommittee on Benefits and Costs, 1950). While never fully accepted either by the parent committee or by the federal agencies, this government report was remarkably sophisticated in its use of economic analysis; the intellectual foundation that it laid for research and debate in the water resources area set it apart from other major reports in the realm of public expenditure. It was fondly known by a generation of resource economists as the "Greenbook."

By the time this codification was published, the evaluation of "conventional" outputs of water resource projects had become routine. They consisted of irrigation, navigation, flood control, hydropower, and municipal and industrial water supplies. A common feature of all of them was that benefits could be satisfactorily evaluated by ingenious applications of information generated by markets. Eckstein's *Water Resources Development* was an exposition and critique of these methods and an interpretation of them in terms of formal welfare economics.

The Integrated River Basin Development Era

Eckstein's classic study stood, as implied, on the brink of a new approach to analysis in water resources economics. One major departure from the past that characterized this new regime was a shift from the focus on individual projects and their evaluation to concentration on the development of river basins as systems, with explicit emphasis on the physical interdependence among the units within such systems. For example, regulation of a river by one impoundment for power production may increase the yield of power at downstream hydro units because they receive a more even river flow. The other major departure was that, while previous research had focused entirely on issues concerning water quantity, the economics of water quality began to receive serious research attention during the 1960s, as discussed below.

In retrospect, one can see that *Multiple Purpose River Development* stood in the vanguard of an era of research on the applications of systems analysis to the economics of water resource development. The study was the first to provide a detailed description of the kinds of physical interdependencies that exist in river basins and which must be taken into account if efficient development is to take place. It went on to apply this analysis in case studies of three major river basins in the United States: the Hells Canyon Reach of the Snake River, the Alabama–Coosa, and the Willamette.

The apogee of this line of research was, however, the Harvard Water Program. Both John Krutilla and I had some connections with it, and anyone not having had this experience would have difficulty understanding how exhilarating it was. Electronic digital computing technology was still relatively new, and the enthusiasm for applications of systems analysis, though in retrospect naive, was almost boundless. These were heady times indeed. The key words can conjure them up—stochastic hydrology, mathematical programming, system simulations, Lagrangian analysis, decision theory. The culmination of this effort was the publication of *Design of Water Resource Systems* by Arthur Maass and associates (1962).

In the same tradition (although done with fewer of the high-tech methods of the time) was the work by Krutilla (1967a) on the Columbia Basin. In it he studied the economics of existing and proposed reservoir and power plant structures in the basin in light of the treaty negotiations then in progress. He analyzed the economic gains possible from developing the Columbia in a systematic way jointly with Canada, studied various possible international distributions of such gains, and illuminated strategic elements of international negotiations. It is the most complete and careful economic analysis of an international treaty concerning natural resources ever done.

Some of the methodologies (e.g., stochastic hydrology and mathematical programming) from the period of research that focused on integrated river basin development have continued to find applications. But the great edifice of economic systems analysis erected by the Harvard Water Program has stood largely empty since the *Design of Water Resource Systems*. This for two main reasons. First, although the physical opportunities existed, integrated river basin development proved to be a rare thing, even to the limited extent that it took place in the Columbia Basin. Much more typical is a disjointed process, with a project being planned, possibly authorized by Congress, and then later, often many years later, money possibly being authorized by Congress for construction, which, again, might be strung out over many years. Indeed, for many authorized projects money was never appropriated. River basin development simply has not been based on system optimization. Second, the studies discussed here proceeded during what was perhaps the peak of dam building in the United States. But at the same time that era was already drawing to a close. Most of the good, or acceptable, sites for water projects had already been developed or were under active development. The attention of economists was beginning to be directed toward other aspects of water resources problems.

The Water Quality Era

Perhaps the major indicator of the reorientation of water resources research was the emerging perception that water quality problems might be just as worthy of economists' attention as the matter of water quantity. This development was in no small measure due to the insight and efforts of John Krutilla and his then codirector in the RFF Water Resources Program, Irving K. Fox.

In early 1960, they hired me to develop a program of water quality studies for RFF. Both of them were extremely helpful and encouraging during my opening struggles, and in the early 1960s a program was in fact launched. The first major publication from this effort was *The Economics of Regional Water Quality Management* (Kneese, 1964).

By example as well as by word, John taught me a lot about how to conceptualize a research program in operational terms. Furthermore, he stressed how one could use RFF's unique situation—the organization having both a strong internal research staff and a grant program that could involve the best outside scholars in the research—to mount a concentrated attack on a research area.

The Water Quality Program was very active in the early through the middle 1960s, focusing on such matters as alternative policy instruments (e.g., effluent charges versus direct controls), methods for modeling the economics of regional water quality management, and institutions for water quality management. This research was instrumental in influencing water quality policy in several countries. A comprehensive report on the program is found in the RFF volume *Managing Water Quality: Economics, Technology, Institutions* (Kneese and Bower, 1968).

The Fallow 1970s

In the 1970s, water resources research as such suffered a considerable decline in interest on the part of the economics profession. Theoretical work on effluent charges and other policy alternatives continued, as did work on the enduring theme of the economics of water allocation in arid areas. And agricultural economists continued to work on various issues. But the tremendous blossoming of interest in this field that had characterized the 1950s and 1960s faded.

In economic research, one theme of the seventies was that water was often seen in the perspective of a wider environmental or resource concern. For example, in the RFF Quality of the Environment Program, water quality was treated as one player in an integrated residuals management approach. There was heavy emphasis on trade-offs among residuals streams that could be discharged to alternative environmental media. The major case study performed during this period (Spofford, Russell, and Kelly, 1976), dealing with the lower Delaware valley, treated water quality issues in a model that handled land disposal, wastewater effluents, and atmospheric emissions at the same time.

Two major case studies performed in arid and semiarid regions looked at water as a possible constraint on economic development and dealt with allocation and reallocation issues. One of these studied the Four Corners region— Arizona, Utah, Colorado, and New Mexico—of the United States and is reported in *The Southwest Under Stress: National Resource Development Issues in a Regional Setting* (Kneese and Brown, 1981). The other addressed concern about the possible inhibiting effect of groundwater depletion on economic growth in the six-state, Ogallala aquifer region (High Plains Associates and coauthors, 1982).

Late 1970s to Mid-1980s: Focus on Measuring Benefits

The project evaluation era as epitomized in Eckstein's *Water Resources Development* gave much attention to the evaluation of benefits. But, as implied earlier, most of the valuation problems involved the production of "private" goods, even though they were supplied by public water resource projects. For these, information generated by markets could be used fairly directly to establish values, although the "alternative cost method" of doing so was subject to, and often experienced, misuse. There was, however, one important "public good" (in the sense of being characterized by jointness in supply) produced by such projects that received much attention in this early period of the application of economics to water resource matters. That good was flood control. By the time of Eckstein's book, an ingenious method, based on the technique of discounting the expected value of losses, was well developed and routinely used. The behavioral assumptions underlying this method, which we would now say was based on expected utility theory, were, however, regarded by some as being questionable (Kates, 1962) (foreshadowing debate over benefits-estimation techniques that has been revived recently).

With an important exception, mentioned shortly, the system analysis and regional water quality eras saw a shift of attention to supply-side concerns. While the importance of benefits issues was clearly recognized, the matter of attaining efficiency on the supply side became the focus of most research during those years. For example, as mentioned, a central concern of *Multiple Purpose River Development* was the external effects of the development of projects at one location in a river system on the efficiency of development at other locations. Much of the work on water quality concerned itself with the problem of meeting externally fixed ambient standards in an efficient manner. An important early example was work done at RFF by Robert K. Davis and reported in his 1968 study, *The Range of Choice in Water Management: A Study of Dissolved Oxygen in the Potomac Estuary*.

The major exception to this general refocusing of research attention was the evaluation of recreation benefits, which was developing actively in the 1960s. Much of the early work in this area had to do with the evaluation of water-based recreation benefits from new reservoirs. The seminal work—*Economics of Outdoor Recreation*—underlying this activity was by Marion Clawson and Jack L. Knetsch (1966), both then working at RFF. There were also, however, a few major studies of the evaluation of benefits resulting from water quality improvement, such as that by Davidson, Adams, and Seneca (1966) on the Delaware estuary.

Like other economics research on water resources, the work on recreation evaluation declined markedly in the late sixties and in the seventies. But a combination of factors brought on intense activity in this area by economists in the late seventies, and it continues today. It focuses on developing methods and data for measuring the benefits of environmental improvement, including improvements in water quality. One reason for this development was that the great cost of environmental improvement programs was becoming apparent. At the same time, the nation's macroeconomic performance was weak. These factors, and perhaps others, prompted the Environmental Protection Agency to support a major program of research on environmental benefits, including those from water quality improvement.

This work differed from much of the earlier work on benefit evaluation in that it dealt almost entirely with public goods issues, including difficult questions associated with effects on health, recreation, aesthetics, and even broader national or regional values (intrinsic benefits). A nontechnical survey of this work through the early 1980s is presented in a book from RFF, *Measuring the Benefits of Clean Air and Water* (Kneese, 1984).

Once again, an important element in laying the conceptual foundation and in defining the issues for this newer work on benefits estimation was a classic by John Krutilla (1967b), together with the RFF program of research on natural environments that he based on it. In that work, a paper entitled "Conservation Reconsidered" (reprinted in the appendix in this volume), he defined and stressed the importance of both use and nonuse values for preservation decisions. He also showed the relevance of concepts such as "option values." All of these concepts have played an important role in the benefits research of the late seventies and early eighties. Other essays in this book concern themselves more centrally with John's work on natural environments; I turn now to my personal view of what lies ahead for economic research on water resources issues.

What of the Future?

Water resources research has always been driven by the perception of problems and the provision of funding by sponsors. The fact that public concern, or at least media concern, with water problems is once again high is amply attested to by the spate of magazine and newspaper articles and television programs that have appeared in the last several years. If this concern results in greater availability of money for research, I think the attention economists pay to water research issues will also, once again, rise. What are the areas ripe for increased attention? I think that two will stand at the center in the waning years of the eighties and opening years of the nineties. They are broader in scope than the benefits research now being carried out, but nevertheless they encompass benefits issues in a very significant way.

The first is water allocation in the arid West. While this has been an enduring theme in water resources research, it has taken on new urgency. With the impending completion of the Central Arizona Project and the Central Utah Project, the water development era in the West will essentially be over. Import of water to the region on a major scale is grossly uneconomic.

But water allocation institutions in the West were built up during a period of rapid development of water resources and were designed to foster and aid that development. Prior appropriation law helped to provide a degree of security of supply so that private developers of irrigation had the incentive to commit capital and labor to the construction of diversion and distribution systems. But the relationship between law, institutions, and economic development did not stop there. Even after water courses were fully appropriated, agriculture continued to expand by developing supplies through the federal program to reclaim the arid West. The era of reclaiming arid lands began in the late nineteenth century, and the Reclamation Act of 1902 established this objective as a national goal. The period following the Reclamation Act was one of heavily subsidized and increasingly centralized, large-scale irrigation projects. Long-term, interest-free financing based on "ability to pay" further institutionalized the notion that unappropriated and undeveloped water was itself free, its only cost being the capital cost of constructing works and the subsequent operation and maintenance cost. The Bureau of Reclamation provided dams and diversion works on most major waterways in the West. Transbasin diversion projects were also commonplace. Accompanying the reclamation program was the creation of special water districts as entities responsible for repayment, operation, and maintenance functions with respect to these public works. If projects experienced hardships, contract obligations were deferred. Regions having political clout in Congress were usually treated most generously. Frequently, hydroelectric power production was part of these projects and was, along with funds generated by the sale of federal lands, used as the "cash register" to subsidize irrigation water development.

But, as indicated, the West is rapidly undergoing a major transformation with respect to water. Increasing water scarcity already has brought changes in western development and water use patterns, and much greater changes are likely in the future. In particular, the expansion of irrigated agriculture based on the availability of inexpensive water is ending.

Indeed, once the Upper Basin allotment under the Colorado Compact is developed, the Southwest, including the Four Corners states, Texas, and California, will have to adjust to a gradually declining overall water supply because of the groundwater depletion that is occurring all over the region and to an especially severe degree in Texas and Arizona.

Competition and conflict are inherent in this situation, and all the established institutions for water allocation in the West are being challenged and transformed. Several recent court decisions have called into question state sovereignty over "their" water resources and have declared water to be an item of interstate commerce. Compact allocations are being called into question. Indian water claims threaten the integrity of the prior appropriations system. At the same time, interest in the value of instream uses of water has been rising. The role of market and nonmarket values and institutions and designs for their improved functioning will, I feel sure, be a center of attention for natural resource economists, especially in the West, in the years ahead.

The second and more universal problem is groundwater contamination. In a report to Congress the Office of Technology Assessment (1984) states that while in the aggregate only a small portion of the nation's groundwater is contaminated, the pollution is often near populated areas; furthermore, groundwater is increasingly relied on for a variety of uses. Approximately 50 percent of the nation's drinking supply comes from the ground. Groundwater use has been growing faster than population over the past twenty-five years. It is thus likely that a growing proportion of the population is being exposed to groundwater contaminants.

Even at extremely low concentrations, many toxic organic chemicals pose serious and irreversible health risks. In almost every case of groundwater contamination that has been studied to date, only a few of several hundred possible compounds were actually tested for, and then only after contamination was suspected. But in many of these cases, well water was found to contain concentrations above, and often several orders of magnitude higher than, those commonly encountered in raw or treated drinking water drawn from badly contaminated surface-water sources.

For given contaminant loadings, the extent of groundwater contamination is determined by the hydrogeologic attributes of the aquifer. Groundwater flow is governed by the hydrostatic pressure gradient, by recharge and discharge rates, and by the aquifer's permeability. Contaminants are transported by advection and diffusion along with the slow flow of groundwater. In that oxygen-poor environment, chemical or physical processes of contaminant degradation proceed very slowly. Thus the contaminant plume may move great distances with very little change in toxicity levels and may eventually reach drinking-water wells.

Among the principal sources of groundwater contamination are waste disposal landfills and impoundments, accidental spills, and abandoned oil and gas wells. Stricter regulation of the disposal of potential contaminants in other environmental media, particularly in air and surface waters, and the consequent rising costs of such disposal are likely to increase the pressure for land disposal, much of which is already illegal. It is thus likely that a growing portion of the population will be exposed to contaminants in groundwater.

Concern of the public and of public officials with groundwater contamination is justified not only because of what has happened already, but also, perhaps even more so, because of the "time bomb" aspects of the problem. But it is expensive to prevent groundwater contamination and especially so to carry out cleanup work after contamination has occurred. Given the already existing demands for both public and private resources, it is important that groundwater management be efficient as well as effective.

The area is rife with problems for economists. How can incentives be arranged so as to "optimally" reduce the generation of hazardous materials without further encouraging illegal dumping? How can and how should economic incentives be used to solve this country's thus far intractable problem of siting regional disposal facilities? What kind of institutional arrangements are needed for efficient and effective disposal, and how are such institutions to be governed? How does one evaluate actual and potential (usually small) risks to health resulting from groundwater contamination in view of the fact that experimental evidence on human behavior suggests that expected utility theory is not a satisfactory conceptual foundation?

Not only do these problems await solutions, but another, related challenge lies ahead. Water resources research, particularly where the focus is on economics, is an inherently interdisciplinary pursuit, which is an important part of its appeal to many in this field. In the system integration and water quality phases of the research, bridges were built to the disciplines of engineering, applied mathematics, ecology, and other natural sciences. These bridges are important even today. The range of problems just identified requires an equally creative, but probably more difficult, reaching out to psychology, social psychology, sociology, and collective choice theory. A challenge indeed!

REFERENCES

Clawson, Marion, and Jack L. Knetsch. 1966. *Economics of Outdoor Recreation* (Baltimore, Md., Johns Hopkins Press for Resources for the Future).

Davidson, Paul, F. Gerard Adams, and Joseph Seneca. 1966. "The Social Value of Water Recreational Facilities Resulting from an Improvement in Water Quality: The Delaware Estuary," in Allen V. Kneese and Stephen C. Smith, eds., *Water Research* (Baltimore, Md., Johns Hopkins Press for Resources for the Future).

Davis, Robert K. 1968. *The Range of Choice in Water Management: A Study of Dissolved Oxygen in the Potomac Estuary* (Baltimore, Md., Johns Hopkins Press for Resources for the Future).

Eckstein, Otto. 1958. *Water Resources Development: The Economics of Project Evaluation* (Cambridge, Mass., Harvard University Press).

Federal Interagency River Basin Committee, Subcommittee on Benefits and Costs. 1950. *Proposed Practices for Economic Analysis of River Basin Projects* (Washington, D.C.).

High Plains Associates; Camp, Dresser, and McKee, Inc.; Black and Veatch; and Arthur D. Little, Inc. 1982. *Six-State High Plains Ogallala Aquifer Regional Resources Study*. Report to the U.S. Department of Commerce at the High Plains Study Council.

Kates, Robert William. 1962. *Hazard and Perception in Flood Plain Management Choices*. Department of Geography Research Paper No. 78 (Chicago, University of Chicago Press).

Kneese, Allen V. 1964. *The Economics of Regional Water Quality Management* (Baltimore, Md., Johns Hopkins Press for Resources for the Future).

———. 1984. *Measuring the Benefits of Clean Air and Water* (Washington, D.C., Resources for the Future).

———, and Blair T. Bower. 1968. *Managing Water Quality: Economics, Technology, Institutions* (Baltimore, Md., Johns Hopkins Press for Resources for the Future).

_____, and F. Lee Brown. 1981. *The Southwest Under Stress: National Resource Development Issues in a Regional Setting* (Baltimore, Md., Johns Hopkins University Press for Resources for the Future).

Krutilla, John V. 1967a. *The Columbia River Treaty: The Economics of an International River Basin Development* (Baltimore, Md., Johns Hopkins Press for Resources for the Future).

_____. 1967b. "Conservation Reconsidered," *American Economic Review* vol. 57 (September) pp. 777–786 (Resources for the Future Reprint 67).

_____, and Otto Eckstein. 1958. *Multiple Purpose River Development: Studies in Applied Economic Analysis* (Baltimore, Md., Johns Hopkins Press for Resources for the Future).

Maass, Arthur, Maynard M. Hufschmidt, Robert Dorfman, Harold A. Thomas, Jr., Stephen A. Marglin, and Gordon Maskew Fair. 1962. *Design of Water Resource Systems* (Cambridge, Mass., Harvard University Press).

Office of Technology Assessment. 1984. *Protecting the Nation's Groundwater from Contamination.* OTA-0-233 (Washington, D.C., Government Printing Office).

Spofford, Walter O., Jr., Clifford S. Russell, and Robert A. Kelly. 1976. *Environmental Quality Management: An Application to the Lower Delaware Valley.* Research Paper R-1 (Washington, D.C., Resources for the Future).

Wollman, Nathaniel. 1962. *The Value of Water in Alternative Uses: With Special Application to Water Use in the San Juan and Rio Grande Basins of New Mexico.* Study conducted by Special Committee under the direction of Nathaniel Wollman (Albuquerque, University of New Mexico Press).

PART 2

RESOURCE PROBLEMS AND THEORETICAL ISSUES IN APPLIED WELFARE ECONOMICS

KEY ASPECTS OF SPECIES EXTINCTION: HABITAT LOSS AND OVEREXPLOITATION

Anthony C. Fisher

This paper explores the uses of economic theory in understanding what many noneconomists regard as the major environmental problem of our time—the mass extinction of plant and animal species that is expected in the years ahead. Among economists it is John Krutilla who recognized the problem, along with the closely related issue of wilderness preservation, and who engaged the interest and talents of the profession in addressing it. His seminal work, "Conservation Reconsidered," which appeared in the *American Economic Review* now more than twenty years ago (Krutilla, 1967; reprinted in the appendix in this volume), not only drew our attention to the problem but first showed how concepts of economic theory could help explain it and even offer solutions. This brilliant and profound work marked the beginning of a research program, originally at Resources for the Future (RFF)—the Natural Environments Program—and now represented by a variety of efforts there and elsewhere. It was my privilege in 1975, about halfway through the RFF program, to participate with him in a summing up of what might be regarded as the first phase of the program (Krutilla and Fisher, 1975). In the present paper, I offer two models, drawing on more recent work by myself and others, to describe key aspects of the problem of species extinction: habitat loss and overexploitation.[1]

Although some part of this problem is surely due to the harvesting or overharvesting of particular species, most knowledgeable people believe loss of habitat to be of greater consequence. As the well-known biologist Edward O. Wilson (1980, p. 21) has stated: "The one process ongoing in the 1980's that will take millions of years to correct is the loss of genetic and species diversity *by the destruction of natural habitats*" (emphasis added). This destruction can take several forms: direct conversion, which occurs, for example, with the drainage of wetlands or the development of drylands for agriculture, housing, and transportation; chemical pollution, such as that caused by acid rain; and "biological pollution"—the introduction of exotic species. The major cause of habitat destruction currently appears to be direct conversion for agricultural and other development.

Prevention of widespread extinctions is intimately related to preservation of the natural environment because wildlands and the natural populations that they support are the results of geomorphologic and biologic processes taking place over millions of years, and, if destroyed, these plant and animal communities cannot be replaced or restored. In other words, their loss is irreversible. This of itself is enough to suggest the need for care in making decisions that will lead to loss of habitat. But there is another complicating factor at work here: uncertainty about the value of what will be lost. It is simply not true that every disputed tract is home to a plant or animal species that will prove to be of value to human beings. On the other hand, we can be fairly sure—though only in a statistical sense—that there will be some loss of value resulting from the loss of large tracts, especially of tropical rain forests that support a rich variety of species (many of which are as yet undiscovered and therefore untested for useful properties). Further, information on the nature of the values at stake is likely to improve over time as research activities uncover new species, screen them for medicinal properties, explore their uses in industrial processes, crossbreed key characteristics into domesticated crops and livestock, and so forth.

The question explored here is how to evaluate whether or not to develop a natural environment, taking account of both the irreversibility of development and the accumulation over time of information about the values (if any) that development would preclude. The following section presents a model to describe the development decision, and the next section a model of how the efficient harvesting path to extinction for a renewable resource is modified by recognizing the value of the unharvested stock as a store of value. The last section makes some concluding observations.

A Model of Habitat Loss

The decision to be made is whether or not to develop a tract of wildland. I assume it will be made on the basis of an evaluation of benefits and costs, with the decision rule as follows: develop if the net present value of benefits exceeds that of costs, including the opportunity costs, that is, the forgone benefits of preservation. I assume further that development is irreversible: if the tract is developed in the first of two periods considered, it cannot be preserved in the second period. Finally, I assume that second-period benefits of development and preservation are uncertain but that information will improve and the uncertainty will be resolved by the start of the second period.[2]

In approaching this problem it is useful first to consider what happens when the decision maker deals with the uncertainty about second-period benefits by simply replacing random variables with their expected values, a common practice in applied analysis. The resulting assessment of the development project and the corresponding decision are then contrasted with the assessment and decision that would follow a proper accounting for the prospect of new information, and the implicit assumption of risk neutrality in both cases is further discussed.

Suppose, then, the decision maker chooses to preserve or to develop in the first period on the basis of known first-period and expected second-period benefits. Net present value over both periods, as a function of the first-period decision (0 or 1, where 0 = preservation and 1 = development), is

$$V^*(0) = B_1(0) + \delta \max \{E[B_2(0, \theta)], E[B_2(1, \theta)]\} \tag{1}$$

or

$$V^*(1) = B_1(1) + \delta E[B_2(1, \theta)] \tag{2}$$

where B_1 = first-period benefit
  B_2 = second-period benefit,
  δ = a discount factor,
  θ = a random variable.

Note that, as a consequence of the irreversibility assumption, first-period development is locked in (equation (2)).

The development decision rule is

$$d_1^* = \begin{cases} 0 & \text{if} \quad V^*(0) - V^*(1) \geq 0 \\ 1 & \text{if} \quad V^*(0) - V^*(1) < 0 \end{cases} \tag{3}$$

where d_1 = first-period development.

Observe that if $E[B_2(1, \theta)] \geq E[B_2(0, \theta)]$, the current development decision is based solely on a comparison of current preservation and development benefits, $B_1(0)$ and $B_1(1)$.

Now suppose the first-period decision allows for new information that will resolve the uncertainty about second-period benefits by the start of the second period. Net present value in this scenario is given by

$$\hat{V}(0) = B_1(0) + \delta E\{\max [B_2(0, \theta), B_2(1, \theta)]\} \tag{4}$$

or

$$\hat{V}(1) = B_1(1) + \delta E[B_2(1, \theta)]. \tag{5}$$

Note that second-period benefits are not replaced by their expected values. Instead, the decision maker is assumed to learn which second-period option, $d_2 = 0$ or $d_2 = 1$, will yield greater benefits and to choose that one. Of course, at the start of the first period, when d_1 must be chosen, the decision maker has only an expectation as to which will prove greater.

The development decision rule is

$$\hat{d}_1 = \begin{cases} 0 & \text{if} \quad \hat{V}(0) - \hat{V}(1) \geq 0 \\ 1 & \text{if} \quad \hat{V}(0) - \hat{V}(1) < 0. \end{cases} \tag{6}$$

What can be said about value-maximizing or optimal development in the first period in each case? Clearly, since $V^*(d_1)$ and $\hat{V}(d_1)$ are different, d_1^* and $\hat{d}_1$ will be different. A natural hypothesis is that $\hat{d}_1 \leq d_1^*$, since it would seem to make sense to put off development, which is irreversible, if there is a prospect of better information about the benefits it will preclude. Stated differently, if the decision maker ignores the prospect of better information and simply replaces random variables with their expected values, first-period development will be too great. This is precisely the result we obtain. From the convexity of the maximum operator and Jensen's inequality,

$$\hat{V}(0) - V^*(0) = \delta E \{\max [B_2(0, \theta), B_2(1, \theta)]\}$$

$$- \delta \max \{E[B_2(0, \theta)], E[B_2(1, \theta)]\} \geq 0. \tag{7}$$

Since $\hat{V}(1) = V^*(1)$, it follows from (7) that $\hat{d}_1 \leq d_1^*$. This means that optimal first-period use of the tract is *less* likely to be full development ($d_1 = 1$) when the decision takes proper account of the prospect of new information. Conversely, ignoring this prospect and replacing random variables with their expected values (an easy trap to fall into) will bias the decision in favor of development.

Qualifications and Extensions

Two implicit assumptions that underlie the result as stated merit some discussion here. One involves treatment of the decision maker's attitude toward risk: in working with expected values, I have implicitly assumed a risk-neutral attitude. Following the arguments of Arrow and Lind (1970) and others, this may be appropriate in a social decision. But even if it is not, note that the result is independent of any assumption about risk preferences. Although benefits might normally be measured in money units in the setting of an applied analysis, nothing in our formulation requires this. Benefits could just as well be measured in utility units, in which case the decision maker is maximizing expected utility and displays risk aversion. The result—a traditional bias in favor of development, or habitat loss—continues to hold. Of course, the numbers in an actual application would be different depending on whether or not a risk premium of some sort is added to the quantity $\hat{V}(0) - V^*(0)$ in equation (7). This might represent a difference between private and social benefit evaluation if the private evaluation cannot properly adopt a neutral attitude toward risk. There may be other differences as well. For example, some information may not have private value. But in any case (risk neutral or risk averse, public or private), the *qualitative* result in equation (7) holds.

A second assumption, which may be troubling to some, is that the decision maker obtains perfect information, as opposed to merely better information, about second-period benefits. This assumption is made solely to simplify the notation. Elsewhere, Hanemann and I show that an analogous kind of result goes through in a more standard Bayesian setting in which information is updated from period to period but does not completely resolve uncertainty (Fisher and Hanemann, 1986).

Finally, an explicit assumption: resolution (in the present model) is independent of the first-period development decision. This is clearly an empirical matter. One can imagine a situation in which undertaking some development in the present yields information about future benefits. But where the uncertainty is largely about future benefits of habitat preservation—what will be found, what might it be good for, and so on—resolution will come (if at all) not from destruction of habitat but from research into the nature and properties of the indigenous species. The research may be stimulated by consideration of a tract for development, but the information flows from the research and not from the act of development.

Also, note that in the development-dependent learning scenario, it must be possible to develop a part of the tract in such a way that information is provided about the rest without, at the same time, affecting it. The problem here is that if the development is not to foreclose substantial potential future benefits by locking in a choice that may prove mistaken, it must be "small." But, then, it can be shown that a result much like the present one—a bias in favor of full development over small development—follows if the prospect of information from the small development is neglected (Fisher and Hanemann, 1987). In the present case, where the choice is whether or not to develop the tract in the first period, the concept of dependent learning is simply not relevant unless one accepts as a rational argument something analogous to the celebrated explanation from the Vietnam era: "It was necessary to destroy the village in order to save it."

A Model of Overexploitation

Much of the economic literature on the possibility of extinction of a renewable resource, such as a fishery, has focused not on habitat destruction but on the rate of harvest (see, for example, Berck [1979] and Clark [1976]). It is well known that the common property nature of some of these resources, again exemplified by the fishery, leads to overexploitation, since more of the resource is harvested sooner than would be the case under sole ownership. Consequently, extinction is more likely. But the work of Clark, Berck, and others goes deeper. It suggests that even under sole ownership a resource can be optimally exploited to the point of extinction. This section presents a simple model of renewable resource harvest that confirms this result. It then shows how less orthodox notions of resource value, introduced and emphasized by John Krutilla, affect the result.

Since the notion of a steady state is crucial to the analysis, a multiperiod decision framework is needed instead of the simple two-period framework of the preceding section. I continue to assume, however, that the problem faced by the decision maker is to allocate a resource efficiently over time. I also continue to assume that the decision in question is public rather than private, which is appropriate for two reasons. First, I wish to abstract from the open access problems caused with common property resources in situations involving many private harvesters. Second, in a sequel, I introduce an element of public good benefit from the unharvested stock. To put the matter a bit differently, the resulting (efficient) allocation would not ordinarily be achieved by the private sector both because it would be more subject to common property conflicts and because it would not capture any public good benefits.

One other assumption is made which differs from that in the preceding section: uncertainty is not explicitly considered. Clearly the benefits of harvesting or of not harvesting are uncertain. However, the arguments put forward in this section do not depend on uncertainty. Accordingly, the model presented here is not, it should be emphasized, an extension of the preceding model. Instead, it is a different approach featuring different analytical elements to deal with a different concern: overexploitation rather than habitat loss.

The decision problem can be stated as one of maximizing the present value of benefits from the resource where benefits can be given the standard interpretation of combined consumer and producer surplus. In symbols, the problem is

$$\max_{y_0, y_1, \ldots, y_{T-1}} \sum_{t=0}^{T} \frac{\int_0^{y_t} p(z) \, dz - c(y_t)}{(1+r)^t} \tag{8}$$

subject to

$$X_{t+1} - X_t = g(X_t) - y_t \tag{9}$$

where y_t = extraction in period t,
$\quad X_t$ = the resource stock in t,
$\quad p(\cdot)$ = inverse demand function for the resource,
$\quad c(\cdot)$ = (total) extraction cost,
$\quad r$ = (social) discount rate,
$\quad z$ = variable of integration,
$\quad g(\cdot)$ = growth of the stock.

Solution of the problem yields

$$\frac{\lambda_t - \lambda_{t-1}}{\lambda_{t-1}} = r - \left(\frac{\lambda_t}{\lambda_{t-1}}\right)\frac{dg}{dX_t} \tag{10}$$

where λ is a Lagrange multiplier, interpreted as the shadow price of a unit of the resource in the stock. In a steady state the shadow price is not changing, so equation (10) becomes simply

$$\frac{dg}{dX_t} = r, \tag{11}$$

which can be represented as in figure 1. But note that in another possible outcome the steady-state stock is not $X = X^*$ but $X = 0$. If the natural productivity of the species, dg/dX, remains below the productivity of capital in the economy, r, even as the stock dwindles, the two curves can fail to intersect at any positive X. This possibility is represented by the dashed-line curve in figure 1. (Note that subscript "t" is dropped for steady state.)

Thus, extinction of the resource can occur even in the absence of common property externalities. The extinction can be rational if the objective is to maximize the net present value of the harvested resource product. Given a positive rate of discount, it does not pay to wait for a slowly growing natural population to regenerate itself.

However, as suggested elsewhere in the context of multiple use forest resources (Bowes and Krutilla, 1985; Hartman, 1976), the unharvested stock can be regarded as a store of value: genetic information that can lead to applications in medicine, agriculture, and other areas. If it is further assumed that this nonextractive value depends (positively) on the size of the stock, the result quickly follows (as shown below) that the optimal steady-state stock is increased. This assumption can be questioned. Might it not instead be true

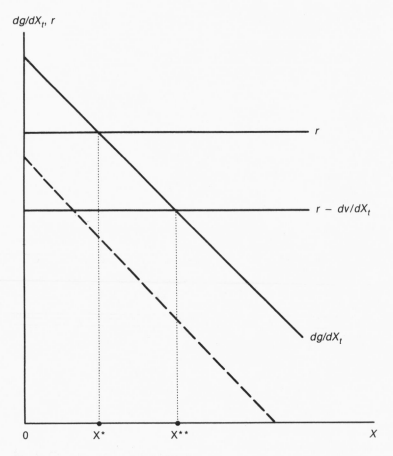

Figure 1. Bioeconomic equilibria.

that as long as a number of individuals remain, sufficient to reproduce, merely adding to the number provides no benefit? It turns out that this is *not* true. Variability among different populations of the same species and within populations is crucial to the ability of species to successfully evolve in response to changes in their environment and, more to the point, to be useful in a variety of applications—and variety depends on the size of the stock. For example, wheat in the American Northwest loses its resistance to rust within about five years. New strains must be developed from infusions of wild varieties (Ehrlich, Ehrlich, and Holdren, 1977). Another example is a recently discovered wild-grass relative of corn that is a perennial. If this characteristic can be introduced in the domestic corn crop, the savings from elimination of annual plowing and seeding (and consequent soil erosion) will be very substantial (Vietmyer, 1979).

Let us then change the objective function in equation (8) by introducing a term for value as a function of stock size, $v(X_t)$. The solution, equation (11), becomes

$$\frac{dg}{dX_t} = r - \frac{dv/dX_t}{\lambda_t}. \qquad (11)$$

Since $dv/dX_t > 0$, the second term on the right-hand side is negative, and the optimal stock is increased to X^{**} as indicated in figure 1. Extinction is less likely, although it is still possible.

Conclusions

Plant and animal species—renewable resources of great value—are facing extinction as a result of the activities of "economic man." This paper explains that paradox with the aid of models of habitat conversion and resource exploitation. It also shows how the models are appropriately modified, in ways suggested by the work of John Krutilla, to produce a different outcome—one in which extinction is less likely to appear as rational economic behavior.

Even this may be unsatisfying to one who takes an explicitly noneconomic approach to the problem of extinction. For example, it is certainly possible to argue on religious or ethical grounds that all species are "sacred" or, at least that they have a right to exist which is independent of any value—agricultural, industrial, medicinal, or even aesthetic—they may yield to human beings. This paper does not directly take issue with such a view. It simply explores some implications of the economic approach. Yet, in the hard choices that confront us—how much tropical rain forest to reserve from development that would destroy habitat, where to set up reserves, how to finance them, and so forth— it may not be possible to avoid the economic approach. Some balancing of benefits and costs, some trade-offs, are likely to be required, not by some obscure economist but by the realities of decision making in the national and international political arenas. As such decisions come to be made, those concerned about extinction will find a large measure of support in the economic approach pioneered by John Krutilla.

NOTES

1. For a somewhat different approach and a comment that clarifies the difference between that model and the original Krutilla–Fisher formulation, see Bishop (1978) and Smith and Krutilla (1979), respectively.

2. The model is based on the original formulations by Arrow and Fisher (1974) and Henry (1974) but adopts the more transparent notation of Hanemann (1983). For extensions to many periods and partial instead of perfect information, see Fisher and Hanemann (1986).

REFERENCES

Arrow, K. J., and A. C. Fisher. 1974. "Environmental Preservation, Uncertainty, and Irreversibility," *Quarterly Journal of Economics* vol. 88 (May) pp. 312–319.

Arrow, K. J., and R. C. Lind. 1970. "Uncertainty and the Evaluation of Public Investment Decisions," *American Economic Review* vol. 60 (June) pp. 364–378.

Berck, P. 1979. "Open Access and Extinction," *Econometrica* vol. 47 (July) pp. 877–882.

Bishop, R. C. 1978. "Endangered Species and Uncertainty: The Economics of a Safe Minimum Standard," *American Journal of Agricultural Economics* vol. 57, pp. 10–18.

Bowes, M. D., and J. V. Krutilla. 1985. "Multiple Use Management of Public Forestlands," in A. V. Kneese and J. L. Sweeney, eds., *Handbook of Natural Resource and Energy Economics* vol. II (Amsterdam, North-Holland).

Clark, C. W. 1976. *Mathematical Bioeconomics* (New York, Wiley).

Ehrlich, P. R., A. H. Ehrlich, and J. P. Holdren. 1977. *Ecoscience: Population, Resources, Environment* (San Francisco, W. H. Freeman).

Fisher, A. C., and W. M. Hanemann. 1986. "Information and the Dynamics of Environmental Protection: The Concept of the Critical Period," University of California, Department of Agricultural and Resource Economics. Working Paper No. 420 (Berkeley, June).

Fisher, A. C., and W. M. Hanemann. 1987. "Quasi-Option Value: Some Misconceptions Dispelled," *Journal of Environmental Economics and Management* vol. 14, no. 2, pp. 183–190.

Hanemann, W. M. 1983. "Information and the Concept of Option Value," University of California, Department of Agricultural and Resource Economics. Working Paper No. 228 (Berkeley, November).

Hartman, R. 1976. "The Harvesting Decision When a Standing Forest Has Value," *Economic Inquiry* vol. 14, pp. 52–58.

Henry, C. 1974. "Investment Decisions Under Uncertainty: The Irreversibility Effect," *American Economic Review* vol. 64 (December) pp. 1006–1012.

Krutilla, J. V. 1967. "Conservation Reconsidered," *American Economic Review* vol. 57 (September) pp. 777–786 (Resources for the Future Reprint 67).

———, and A. C. Fisher. 1975. *The Economics of Natural Environments: Studies in the Valuation of Commodity and Amenity Resources* (Baltimore, Md., Johns Hopkins Press for Resources for the Future).

Smith, V. K., and J. V. Krutilla. 1979. "Endangered Species, Irreversibility, and Uncertainty: A Comment," *American Journal of Agricultural Economics* vol. 61 (May) pp. 371–375.

Vietmyer, N. D. 1979. "A Wild Relative May Give Corn Perennial Genes," *Smithsonian* vol. 10, pp. 68–76.

Wilson, E. O. 1980. "Resolutions for the '80s," *Harvard Magazine* (January–February).

INTERGENERATIONAL EQUITY AND THE SOCIAL RATE OF DISCOUNT

Talbot Page

According to adherents of the positivist program, from Robbins to Friedman and others, economists should describe, explain, and predict but not evaluate and prescribe; economics should be positive and not normative. This view (itself prescriptive) of the proper subject of economics clashed with the view adopted earlier by Adam Smith and other classical economists who integrated positive and normative discussions. It clashes as well with the subsequently developed view which is the subject of this volume: the view of applied welfare economists, who, like the classical economists, employ a mixture of positive and normative ideas.

Applied welfare economists developed benefit–cost analysis because they found the needs of practical decision making—specifically, of decision making related to water resources—too important to let themselves be bound completely by positivist thinking. But perhaps influenced by the positivists, applied welfare economists have limited themselves for the most part to one normative idea—efficiency—although often seeing efficiency as being so universally appealing and analytically tractable that they scarcely think of it as a normative idea at all.

Applied welfare economists instituted a division of labor in normative studies—they themselves focusing on efficiency, leaving equity and other normative issues to political philosophers, legislators, and legal scholars—that has proved very productive. A great deal of efficiency-based analysis has been developed, as is illustrated in this volume. But there comes a time when divided labors need to be brought together. There would be little point to Adam Smith's pin factory if the wire, once drawn, were not then cut and shaped into pins. Likewise, equity and other normative concepts besides efficiency must be brought together to complete the normative analysis. Such integration is especially needed when long-term economic effects are under consideration, because dealing with extended periods of time can complicate and obscure our ideas of equity and efficiency.

Broadly, the purpose of this paper is to take a step toward integrating normative concepts of efficiency and equity into a common framework. (It should be noted here that *efficiency* and *Pareto-optimality* are used interchangeably in this paper.) In one sense this is a step farther away from the positivists, although in another sense the normative analysis of this paper is traditional—an intratemporal analog of such a common framework being the familiar Edgeworth box. Such a diagram provides a framework for discussing both efficiency (being on the contract curve) and equity (choosing a fair endowment or point on the contract curve). This paper, then, seeks to provide a framework analogous to an Edgeworth box but one that is suitable for intertemporal, or intergenerational, analysis.

Intergenerational efficiency is closely related to the social rate of discount. Indeed, the two are sometimes confused; that is, it is sometimes thought that intergenerational efficiency requires discounting as a social choice rule. This paper shows, however, that intergenerational efficiency and the social rate of discount are distinct concepts and that efficiency does not require discounting as a social choice rule. I show this by a counterexample, drawn from Arrow's axioms, which also shows that *not* discounting (at the level of social choice) does not mean discounting at a zero rate. For comparison I provide an example, drawn from Koopmans' axioms, where there is discounting as a social choice rule. The relation between discounting and efficiency is untangled by noting that discounting can arise at three levels: (1) at the level of definition of the feasible set, (2) at the level of individual preference orderings, and (3) at the level of social choice rules.

Before beginning the analysis, however, I would like to say a few words in appreciation of John Krutilla and the part he played in my developing these ideas. In 1967 John published "Conservation Reconsidered" (Krutilla, 1967; reprinted in the appendix in this volume). As one looks back over the past twenty years, it is clear that this was a remarkable paper. In it John set out the problem of intergenerational conflicts of interest in terms of option demand, existence value, irreversibility, dynamic formations of changing preferences and opportunity costs, divergences between willingness to pay and willingness to accept, and market failures from lack of appropriability and other externalities. These ideas, sketched in a few short pages, were to be expanded by John and other economists over the years into papers, books, and research programs. Indeed, his paper might have been called "Environmental Economics Prefigured."

John wrote that paper in the mid-1960s as his outline for the research program he planned for a small group of economists in the Natural Environments Program at Resources for the Future (RFF). For a few years I was part of this group, which included many of the authors of the papers in this volume.

In one way or another we all worked on the problem of intergenerational conflicts of interest posed in "Conservation Reconsidered." Most of the others focused on efficiency aspects of the problem. That was the focus of my attention too, at the beginning. At the time much of my attention was directed toward choosing a "correct" social rate of discount—a choice that was viewed primarily as an efficiency problem. But, ideas come at unexpected moments, and sitting at a luncheon seminar one day on quite another subject, I was suddenly struck by the idea that the "discount rate problem" might be analyzed at a more foundational level as an equity problem analogous to the intragenerational case (as written up in *Conservation and Economic Efficiency* [Page, 1977]).

After leaving John's program I worked on other things, but from time to time I have returned to the problem that he posed, and that was raised before him by Frank Ramsey (1928) and A. C. Pigou (1932). Gradually I came to believe that the language of social choice is useful for discussing this problem. This paper is a tribute to John, who, along with Allen V. Kneese and others at RFF, posed problems for me and then turned me loose to work on them. I return here to the questions set out in "Conservation Reconsidered" to pursue the problems of intergenerational equity and the discount rate. I will try to explain why I believe that the concept of intergenerational equity is broader and more foundational than that of choosing the "correct" social rate of discount.

Intergenerational Equity: Definition and Context

The problem of intergenerational equity is simply but broadly stated as follows: decisions are made in the present with consequences extending through time for generations. The issue is to develop an adequate answer to the question: "What is a fair distribution of consequences over generational time?" In this paper, I take *equity* and *fairness in distribution* to be synonymous.

Analogy with Intragenerational Equity

The idea of *inter*generational equity can be shown by analogy with an intuitive concept of *intra*generational equity. Suppose that a decision is made in the present, and, by assumption, the distribution of consequences is limited to individuals in the present. The question of intragenerational equity is this: "What is a fair distribution of consequences?" For the moment, we will pursue the intragenerational case (by assuming that the consequences are limited to the present).

In an ordinary market transaction in which there is voluntary exchange on both sides, equity is not a matter of great concern. Presumably, both parties to the transaction are made better off by it. The question of equity could arise, of course, if one party was made a great deal better off and the other only a little; in such a case we may worry about the fairness in the distribution of resources and opportunities (thence market power) going into the transaction, and from that the fairness in the distribution of benefits from the decision. However, generally speaking, when both parties are made better off there is little concern for the equity of the decision and little attention drawn to the underlying distributions of resource control.

More attention is drawn to the question of equity when there are externalities, since in such cases a decision can lead to direct costs to some individuals or groups. Externalities arise in markets where there is pollution and congestion, and they also arise as a result of group decision making. For example, the government of Thailand may decide to relocate Bangkok because it is sinking. There would be many gainers and many losers from such a decision. Similarly, there would be important distributional questions if the World Bank decided to fund a project building a dam or an agricultural project in which an acutely toxic pesticide was used.

In the rare decision where there are no losers and everyone wins, the situation is like an ordinary market transaction without externalities. There is an actual Pareto improvement and (typically) little concern for equity. There may be some arguing over the division of the gains, which is in part a question of equity, but the issue arises more sharply when there are losers. Rather naturally, where the gains and losses are small the concern for equity is small, and where the gains and losses are larger, and more skewed, the concern for equity is larger. Where there are many decisions, all involving relatively small losses and gains, and where a person's loss from one decision is likely to be offset by a gain on the next, the distributions may seem fair "on average." In such cases it is traditional to make decisions on the basis of potential Pareto improvements in the hope that on average there will be actual Pareto improvements. This approach sets aside the question of fairness in distribution for the single decision in the hope that it will be attained in the aggregate.

Reliance on potential Pareto improvement does not work when a series of decisions systematically favors some and hurts others. Nor does it work when there is a politically powerful group that refuses to bear an asymmetric burden and can force a decision to be restructured with different distributional impacts. Nor when costs from one decision are believed to be incommensurate with the gains from another (as often happens in cases of health or liberty), so that the consequences of one decision are not taken to offset the consequences of another.

So far I have briefly described some of the circumstances in which the problem of equity is more pressing and other circumstances in which it is less so. Indeed, it is easier to talk about the circumstances of equity than of equity itself, and perhaps with good reason, since we may be able to come closest to the concept of equity by focusing on the circumstances in which it comes into play. Clearly equity is not simple equality, and in ordinary conversation people tend to react more strongly to what they consider to be unfair than to what they see as fair. A little more directly, where there are equalities in background conditions—for example, where people are treated as equals, with equal justice under law, equal education, equal health care, and roughly equal initial resources and opportunities—we are more confident in judging outcomes as fair. With such equality in background conditions and without involuntarily imposed external costs, we are likely to judge outcomes equitable even if they are quite unequal.

When a decision leads initially to asymmetric distribution of costs and benefits but compensation is then made and the compensating benefits are commensurate with the costs, there may be more concern for harm to autonomy but still little concern for the equity of the decision. When there is appeal to averaging over many decisions, more concern over questions of equity may begin to arise. But when decisions are large, lumpy, and possibly irreversible and when they also involve such basic goods as health, liberty, and the means of survival and affect the process of valuation itself, questions over the fair distribution of consequences arise most sharply.

To move from the concept of intragenerational equity to that of intergenerational equity we simply broaden our focus to include decisions with effects spanning generations. As before, we ask: "What is a fair distribution of consequences over generational time?" Just as we thought previously about actual and potential Pareto improvements among a reference set of people in the present generation, we can think of actual and potential Pareto improvements for a larger reference set, which includes both present and future generations. Further, the same intuitions about the circumstances which describe when equity is of greater or lesser concern carry over to the intergenerational case. In other words, I take the concept of intergenerational equity to be the same as that of intragenerational equity, but with a longer time perspective and a larger reference set.

Although the concepts of intragenerational and intergenerational equity are the same except for the broader context of the latter, this broadening nevertheless introduces four complications. First, the reference set of individuals for which welfare comparisons are to be made is endogenous. The actual individuals living in a future generation change, depending on decisions made in the present. Parfit (1984) has made much of this self-referencing problem, but I will comment on it here only briefly: in moving from particular individuals of a generation to the generation itself, we are saying that individual identities do not matter. While it is plausible that "reasonable" preferences may be abstracted from particular identities—"reasonable" people prefer more health to less—this abstraction raises problems (particularly for population policy) that will not be analyzed here.

The second complication is that, since future generations are not physically present and able to tell us their values and press their interests, there is a fundamental asymmetry in the distribution of power across generations. There can be no direct bargaining between generations as there can be within a generation. We can harm or benefit the future in ways that the future cannot harm or benefit us. There are corresponding asymmetries in power among individuals in the same generation—for example, in varying abilities to pay for legal counsel in a court case—and concern for fairness may lead to provision of court-appointed counsel. Nonetheless, parties with conflicting interests within the same generation can at least talk to each other. In the intergenerational case, if the present generation is to act at all to protect the interests of future generations, it must do so in a strongly fiduciary role.

The third complication introduced into the problem of equity by inter-generational considerations is one of implementation; it is addressed only briefly here.

The fourth complication, which is the primary focus of this paper, is as follows: since there is separation in time across the succeeding generations, we need to consider how time preference and capital productivity fit in.

Intergenerational Social Choice

To sharpen our notion of intergenerational equity, we begin by noting that markets are often left to make decisions with intergenerational distributional consequences. When a government or society endorses and encourages market institutions, the resulting distributions are also implicitly endorsed. The ac-ceptance of the distributional consequences is done indirectly, at the level of choosing the institutional arrangements, but it is nonetheless done. In traditional planning agencies—for example, where the World Bank or a government agency makes an intervention in an economy—distributional consequences are often faced more explicitly. But in either case, either implicitly for markets or explicitly for planning agencies, the issue of intergenerational equity arises.

In the market setting, distributional decisions are made on the basis of present values and the discounting of future costs and benefits. The traditional approach of planning agencies making decisions with intergenerational con-sequences is to mimic the market. The procedure is first to estimate the alternative streams of costs and benefits across time and then to choose the alternative with the greatest discounted net benefits. The main departure from the market approach is that instead of accepting the market rate of interest there is an attempt to choose a "correct" discount rate that is somehow both efficient and "intergenerationally fair." A great deal of work has gone into trying to figure out what the correct rate of discount is.

While the procedure just described is the conventional approach, economists from Ramsey (1928) and Pigou (1932) to Solow (1974) and Harsanyi (1982) have had misgivings about it. Thus, there has long been nagging doubt about how adequately the traditional approach deals with distributional issues, but these misgivings are often ambiguous.

The language of social choice allows us to phrase this nagging doubt more precisely. It provides vocabulary for describing concepts of equity and discounting, and it also provides a syntax. The language is very restricted, having only a small vocabulary. But the syntax turns out to be rich and complicated. The theorems are the syntax, and here there are nonintuitive results. In some cases the nonintuitive results are no more than explicit revelations of hidden assumptions and interpretations in traditional discussions. In my view, traditional discussions of the discount rate problem have been particularly ambiguous in two areas: in the definition of the reference set over which Pareto comparisons are made and in the levels of discounting. When the traditional statement of the discount rate problem is cast more precisely in social choice language, these ambiguities can be clarified and the nagging doubt about it stated more sharply.

One of the most useful aspects of the language of social choice is that it draws a clean dichotomy between aggregation and maximization. As we shall see, intergenerational equity has more to do with aggregation and discounting procedures more to do with maximization.

The aggregation problem (in the intragenerational context) is defined as follows. We start with a society of n people and a set X of alternatives. A particular $x \in X$ describes a possible social choice (or action or decision alternative). An alternative x is a picture or snapshot that describes what each individual in the society gets. Depending on which x is chosen, there will be different distributional consequences. As different individuals are affected differently by the choice, they may have opposing interests in the choice. The (possibly) conflicting interests are represented by preference orders on the set of alternative choices. (The framework, which views conflicting interests as conflicting preference orders, is perhaps too narrow but nonetheless clarifying.) We write R_i for individual i's preference order. The aggregation problem is to define a "good" aggregation rule which maps the individual orderings into a social ordering.

A little more formally, the ordering idea, on the individual level, is that for any two conceivable states, call them x and y, each i will either prefer x to y (write xP_iy) or prefer y to x, or be indifferent. If i prefers x to y or is indifferent, we write xR_iy.

Similarly, the language assumes there is a social ordering between any two conceivable states: either x is socially better than y and we write xPy, or y is socially better than x and we write yPx, or they are socially equal. Again if x is socially at least as good as y, we write xRy. As we go from private to social orderings we take off the subscript and somehow resolve or harmonize conflicts of interest.

The aggregation problem is to find a suitable aggregation rule F that makes this resolution. We write $F : R_1 \times \ldots \times R_n \to R$, where R is the social ordering. If our notion of a *good* rule includes the idea of a *fair* rule, we have cast the problem of equity into the language of social choice.

The basic problem in social choice is to define an aggregation rule F that resolves conflicting interests among individuals. The traditional approach is to define F as the consequence of a set of axioms that the social choice process is to follow. For F to be normatively appealing, it needs to be based on normatively appealing axioms. Traditionally, part of the appeal of various proposed axioms is that they are based on ideas of fairness—properties of symmetry, anonymity, and neutrality. By building our ideas of fairness into the axioms and thence into the aggregation rule F, we become explicit about what we mean as a fair distribution of consequences from a decision.

One of the most useful and foundational characteristics of axiomatic social choice is that it allows for differing and multiple normative concepts to be brought in at the same level. Thus, in addition to equity or fairness we can build in another normative idea—efficiency—simply by adding a Pareto axiom to the list of axioms F must satisfy.

Then, having specified F, we can apply it. For a particular application there is a particular profile of preferences $(R_1, \ldots, R_n)$, and through F a particular social ordering R. Once a social ordering R is defined, we pick the top element of R and that becomes our decision. (A top element is an element at the top of social ordering: an $x \in X$ such that xRy for all $y \in X$. In general, there may be more than one top element, or none. The transitivity assumption, made later, assures existence of a top element, at least for finite sets of alternatives.)

Specifying F in terms of underlying axioms is the aggregation problem; picking off the top element of R, once defined, is the maximization problem. Of the two conceptual tasks the aggregation problem has turned out to be much harder, although in terms of application the maximization problem has received much more attention. *Public Investment, the Rate of Return, and Optimal Fiscal Policy* by Arrow and Kurz (1970) provides an example of both the dichotomy between aggregation and maximization and the conventional focus of attention. The book is an application of control theory to investment planning. Its first four pages are on the aggregation problem—the problem of how to choose the decision criteria that define the aggregation rule—and the last two *hundred* are on the application of the aggregation rule (which in this case is a discount rate rule). The book concentrates on the maximization problem—on finding the top element under the chosen aggregation rule. This paper inverts the emphasis and looks principally at the aggregation problem, treating the maximization problem as a subordinate matter. Part of the reason aggregation is more difficult than maximization is that the former requires a resolution of the conflict of interests among individuals.

The description of the problem so far has been quite general: the individuals might all be from one generation, making it an intragenerational problem, or they might be from different generations, creating an intergenerational problem. From the social choice point of view, it can be either. And thus, in the social choice point of view, it is easy to carry over our firmer intuitions concerning intragenerational equity to the more complicated case of intergenerational equity.

The intragenerational case is generally cast as a finite problem, with a finite number of individuals. We can disaggregate a particular state $x \in X$ into different views of the state $x = (x_1, \ldots, x_n)$. Here x_i is part of the picture, the part that describes what goes to i, and x is the whole picture, describing what goes to everyone. We also disaggregate $x_i = (x_{i1}, \ldots, x_{ik}, x_{1,k+1}, \ldots, x_{1k'})$ where the first k components describe the private goods that go to i and the remaining components describe public goods that go to i. By the definition of a public good, $(x_{1,k+1}, \ldots, x_{1k'})$ will be the same for all i (indicated by a constant subscript of 1). Public goods are the part of the picture that is the same for each person. It is the private goods that are exchangeable, that can be used for compensation, and that make consumption different from one person to another.

The language provides a compact statement of the traditional planning approach, which incorporates discounting procedures and goes under the name of benefit–cost analysis. In this approach the most desirable normative criterion is that of actual Pareto improvement. By this criterion if xP_iy for all i, then xPy (that is, if everyone prefers x to y, then the social ordering should have x over y). The usual justification is that actual Pareto improvements mean welfare improvements for all and would be supported by unanimous consent. But the principle applies so rarely that it is not very useful in application. Somebody almost always gets hurt by a decision, and, as the traditional argument goes, to insist on actual Pareto improvements would be a counsel of paralysis. So generally we do not use this criterion in application but a variant of it.

The first variant is the criterion of potential Pareto improvement, associated with Hicks and Kaldor, which depends upon a possible redistribution. We define a redistribution as a z for which $\sum_i z_{ij} = 0$ for $j = 1, \ldots, k$. By definition a redistribution is a reshuffling of the *private* goods, where some individuals may get more and some less of a private good but with the total stock of each private good staying the same and with each of the public goods staying the same. Then the (Kaldor–Hicks) criterion of potential Pareto improvement says that if there exists a redistribution z such that $(x + z)P_iy$ for all i, then xPy. If we are comparing two states x and y and if we could modify x by a redistribution z and find that $x + z$ is unanimously preferred to y, then by criterion of potential Pareto improvement the social ordering should be x over y. This idea is the foundation of benefit–cost analysis. (See Feldman [1980] for a discussion of this and other variants of the Pareto principle.)

From a social choice perspective, benefit–cost analysis has foundational problems. The principle of potential Pareto improvement is not asymmetric. There are examples of x being socially preferred to y and y being socially preferred to x. And there are counterexamples where the criterion is not transitive—which means that one can find an x that is socially preferred to y, a y that is socially preferred to z, and an x that is not socially preferred to z. Such intransitivity is troublesome for anyone who wishes to define a social ordering. These two problems, failure of asymmetry and of transitivity, have led economists to conclude that benefit–cost analysis does not have a solid foundation (Feldman, 1980).

The principle of potential Pareto improvement (or Kaldor–Hicks) has been amended as well. The Scitovsky amendment says that xPy if there is a redistribution z that makes $(x + z)P_i\,y$, for all i, but there is no redistribution z' that makes $(y + z')P_i x$, for all i. This somewhat more conservative criterion eliminates the asymmetry problem but not the transitivity problem.

Transitivity fails for another social choice rule that we often use: majority rule. We deal with this failure by creating safeguards. For example, since majority rule is not transitive, whoever controls the agenda has a good chance of controlling the outcome. Put another way, in majority rule the potential problem of "tyranny of the majority" can materialize through agenda manipulation and the exploitation of intransitivities. As a practical matter, with a majority rule voting system there is usually some sort of control on the agenda process. There are constitutional constraints, or "rules of the game." In the case of majority rule, there are constraints to protect the minority from a tyranny of the majority, including limits to agenda manipulation and precluding certain things from being subject to majority decision. With such safeguards majority rule works quite well, which suggests that having an undesirable property such as intransitivity does not mean that a rule should automatically be discarded. So, it may be possible that the aggregation rule of potential Pareto improvement can be made more satisfactory by means of constraints.

In the intergenerational version of benefit–cost analysis and the criterion of potential Pareto improvement, the typical approach is to take a discount rate equal to the opportunity cost of capital and then to maximize a present value using that rate. Analogously, I will suggest that we need to think about constraints or safeguards here as well.

A few modifications are needed in order to develop the analogy between the intragenerational case and the intergenerational case. The first modification takes into account an important difference between the two types of cases. In the intragenerational case there is a definite number of individuals and no natural ordering among them, whereas in the intergenerational case the number of generations is not definite, but there is a natural ordering, counting generations from the present by the natural numbers. To reflect these differences, as the first modification I will take succeeding generations to be ordered by the natural numbers, without a last generation. As a somewhat artificial stylization I take $n < \infty$ in the intragenerational case and $n = \infty$ in the intergenerational case. Doing so allows an axiomatic formulation of a discount rate rule, and it also allows us to use the Arrow axioms in a way that avoids the Arrow paradox.

Another modification is to identify each i as a different generation instead of as an individual. We do this by thinking of the intergenerational problem in two steps. At the first step each generation resolves its own intragenerational aggregation problem. Having done so, each generation has its own aggregation rule and hence its own social preference ordering, which we write R_i. At the second step we attempt to aggregate the $(R_i, \ldots, R_t, \ldots)$. To focus on the problem of intergenerational equity, we abstract from the first step, moving from the individuals of generation t to generation t itself, to concentrate on the second.

Instead of thinking of x as a single snapshot of the whole society, we think of x as a possible movie. Each frame t of the movie is a complete description of what generation t gets. The social choice problem, in the intergenerational context, is to choose an entire movie from among many possible films. Preceding this choice is the more fundamental problem of the selection of the aggregation rule that will guide our choice.

Clearly, the actual people in generation t, their number and identities, depend on decisions made by preceding generations. We do not know their identities or tastes but we are primarily concerned with conflicts of interest that arise at different points in time rather than as a result of differing identities. And, when we consider primary goods such as health and liberty, future generations will probably be much like us in their valuation.

One way of abstracting from individual identities (and endogenously determined ones at that) and focusing on conflicts of interest arising at different points in time is to assume each generation has preference structures that are the same as our own. (Under this premise there are still conflicts, because for the same movie the frame facing one generation differs from the frame facing another generation.) We would then be concerned about equity for people "like us" who happen to live at other times. But while the language of social choice permits us to abstract from individual identities in this way, it does not force us to do so.

With this background, we have the tools of language to say something about discounting and equity. Koopmans (1972) derived an axiomatic basis for representing an individual's preference in terms of discounting. His result, reinterpreted in the intergenerational context, is the traditional discount rate rule used in benefit–cost analysis. Thus, his axiomatic derivation can be interpreted as a normative basis for the traditional discount rate rule. (Discussions of

Koopmans' axioms and their reinterpretation in the context of intergenerational choice can be found in Ferejohn and Page [1978] and Dasgupta and Heal [1979].)

In Koopmans' axiomatic characterization of the discount rate rule, the key concept is *stationarity*. The idea of stationarity is this: suppose that two movie films looked the same in the first frame and then looked different after that. We modify both films by snipping off the first frame of each and throwing it away, moving everything up one frame. Thus, the first generation, which was going to get that first picture, gets the second picture. The second generation, which was going to get the third picture, gets the second picture, and so on. Stationarity says that if we start out with two films x and y, and x is socially preferred to y, then, when we modify the films by moving everything forward one frame, the social ordering is preserved. The modified x is intergenerationally socially preferred to the modified y. Formally, write the two programs or films $x = (x_1, x_2, x_3, \ldots)$ and $y = (x_1, y_2, y_3, \ldots)$ and the two modified films $x' = (x_2, x_3, \ldots)$ and $y' = (y_2, y_3, \ldots)$. Then R satisfies stationarity of $xRy \leftrightarrow x'Ry'$.

Koopmans' theorem says that his set of axioms implies that there is a function $U{:}X \rightarrow (-\infty, \infty)$ and a number α between 0 and 1 such that

$$xRy \Leftrightarrow \sum_{i=1}^{\infty} \alpha^{i-1} U(x_i) > \sum_{i=1}^{\infty} \alpha^{i-1} U(y_i) \tag{1}$$

In other words, under Koopmans' axiom system, including the axiom of stationarity, an intergenerational social ordering R takes the form of a discounted sum of utilities, one for each generation.

The normative appeal of this basis for discounting rests on the normative appeal of the axioms, and this appeal is limited. The stationarity axiom, which plays a central role in Koopmans' derivation, seems unmotivated. It is possible to construct examples where every generation prefers x to y and every generation prefers y' to x'; but even though x' is Pareto inferior to y', under stationarity x' is intergenerationally socially preferred to y'. I return to the question of the normative appeal of the stationarity axiom below, but first let us focus on the main point here: that the language of social choice allows us to talk about discounting on three distinct levels—those of (1) individual preference orderings, (2) the feasible set, and (3) social choice rules.

Three Levels of Discounting

With regard to one level of discounting, recall that we start with individual profiles of preferences: R_1, which states the structure of interests of the first generation; R_2, which states the structure of the interests of the second generation, and so on. In the aggregation problem we are trying to aggregate these preferences (or generational interests) by means of a rule F, to a single social ordering, R. Any aggregation rule $F:R_1 \times \ldots \times R_n \rightarrow R$ involves combining both present and future interests. In the background there is a feasible set X over which we are making choices. I will call a social choice aggregation rule which is in the form of a discount rate rule (equation (1)) *discounting at the level of social choice*.

Second, the concept of opportunity cost of capital is embedded in the definition of the feasible set X, where $X = X_1 \times X_2 \times \ldots$ and where more generally we do not assume $X_t = X_1$. In general, the feasible set at t depends on the choices made previously. To define feasible evolutions of $(X_1, X_2, \ldots)$, we take account of the productivity of capital and its opportunity cost in terms of future precluded programs. The intergenerationally feasible paths x (and only the feasible paths) are put in X. Thus, the concept of discounting as the productivity or opportunity cost of capital is incorporated in the definition of X. I will call this form *discounting at the level of the feasible set*.

Third, the concept of each generation's time preference is embedded in the R_ts, because in general each generation not only can have a preference over the snapshot it sees (or lives), but it also has a preference over each entire movie. In general R_t is defined over X and not just X_t.

The language of social choice allows us to describe the time preference of a generation with limited altruism—a generation that focuses its attention sharply on its own frame and has limited interest in future generations. I will call this form of discounting of future interests *discounting at the level of individual preference*. Social choice language also allows us to describe the time preferences of a perfectly altruistic generation—one that gives equal weight to all the other generations. These and other descriptions of time preference are subsumed in the R_t.

In short, we have discounting at the level of the aggregation rule F, the specification of the social choice rule; discounting at the level of specifying X, the feasible set; and discounting at the level of the R_ts, individual time preference.

The Discount Rate Problem

We can now state the discount rate problem, as it is traditionally viewed, but from a more general perspective. In the traditional view there is some decision to be made: for example, whether or not to undertake a development project. The decision entails a flow of costs and benefits to the future which are to be discounted at some rate. The problem is to find the "best" rate. "Best" might be defined as intergenerationally efficient or fair or as some combination of the two. This best rate is often called the "social rate of discount," or the "correct" rate (Organisation for Economic Cooperation and Development, 1983, pp. 20–21). If the sum of net benefits, discounted by this "social" rate of discount, is positive, the project should be undertaken.

In terms of equation (1), define $U(x_i)$ as the net benefits to generation i with the project (or with an affirmative decision) and $U(y_i)$ the net benefits to generation i without the project (or with a negative decision). Often $U(\cdot)$ is defined as a consumer plus producer surplus. The social rate of discount is represented by the discount factor α.

In traditional analysis the discussion begins with equation (1) as a social choice criterion. There is usually no justification of (1) on the basis of Koopmans' axioms or some other axiom system. The discount rate problem is stated as follows: "Having accepted equation (1) as the intergenerational social choice rule, what is the best α?"

Within the social choice framework, however, there is not just one "correct" discount rate but three distinct concepts of discount rates. Further, there is no single discount rate associated with time preference. Each generation i can have its own time preference built into the specification of R_i. Similarly, there may be no single discount rate associated with productivity, as productivity can vary across generational time.

There is an obvious reason for discounting at the level of individual time preference: respect for "consumer sovereignty" or, as appropriate here, for "generational sovereignty." There is also an obvious reason for discounting at the level of the feasible set: without doing so the intergenerational feasible set would be ill-defined. But there is no obvious reason for discounting at the level of social choice, as the next section shows.

Efficiency and "Intergenerational Majority Rule"

The reason traditionally offered for choosing a discount rate rule for intergenerational choice is to ensure intergenerational efficiency. But what if we could find other social choice rules that did not involve discounting at the level of social choice but which nonetheless were efficient? If that were possible, there would be no reason, on efficiency grounds, for choosing the discount rate rule (at the level of social choice) over the others.

This question is important in the context of this volume because many of the normative criteria are developed as intergenerational efficiency criteria. These include the derivation of efficiency conditions for option demand, existence value, irreversibility, and so on. Once defined, these efficiency conditions are compatible with the Pareto axiom of intergenerational social choice. In other words, these efficiency conditions are subsumed in the Pareto axiom, becoming part of its extensive definitions. This means that if we found other social choice rules which included the Pareto axiom but not discounting at the level of social choice, then the efficiency arguments concerning irreversibility and the like would be no reason for discounting at the level of social choice.

The Arrow axioms, reinterpreted in the intergenerational context, provide just such an example. These axioms are transitivity, Pareto efficiency, independence of irrelevant alternatives, and nondictatorship. (For a discussion of these axioms in an intergenerational context, see Ferejohn and Page [1978]).

Each of Arrow's axioms incorporates a notion of fairness, because in each, each voter, now interpreted as a generation, is treated symmetrically. Arrow's theorem says that when the number of individuals is finite, the four axioms are inconsistent. However, Hansson (1976) has shown that when the number of individuals is infinite all four axioms are consistent. Thus, in the intergenerational context with no last generation, Arrow's impossibility result is avoided.

It turns out that when the number of generations is infinite, there is an infinite number of social choice aggregation rules that are nondictatorial and consistent with the other three Arrow axioms. And they all have a very interesting property: they are all oriented toward the future as opposed to the present. Say that three or four generations prefer x to y, but every generation after that prefers y to x; then the intergenerational social choice is y to x. The Arrow axioms are like an intergenerational majority voting rule where the infinite majority beats the finite minority. Note that this decision rule is time asymmetric, because all the infinite majorities are in the future. And there can only be a finite minority if there is a switch in preferences after a finite number of generations. If the future can be unified against the present, then it becomes a decisive set and it beats the present. While this is an interesting property, its orientation toward the future is so dominating that I am not recommending it as an intergenerational choice rule. (It says, for example, that if the next thousand generations prefer x to y, and then after that all generations prefer y to x, we should pick y to x for the intertemporal social choice.)

The result also has the weakness that it is heavily dependent on the assumption of an infinite number of generations. The result would be more convincing if it did not turn so critically on whether $n = \infty$ or $n < \infty$. Nonetheless, the main idea is to use the social choice language to disentangle concepts of discounting from those of equity. And in fact the "technical" assumption $n = \infty$ works in our favor in this regard, for the discounting rule implied by Koopmans' axioms also requires the assumption $n = \infty$, and thus the two axiom systems are on equal footing, making the comparison more direct.

Compared with Koopmans', Arrow's axioms lead in a different direction with completely different social choice rules, but rules that are also efficient intergenerationally (this is guaranteed by Arrow's Pareto axiom). Arrow's axioms are future-oriented; Koopmans' axiom of stationarity is also time-biased, but toward the present. To see this we put Koopmans' stationarity axiom in the context of Arrow's first three axioms and obtain a striking result. Instead of having an infinite number of social choice rules that are compatible with these four axioms and which are nondictatorial, Ferejohn and Page (1978) have shown that all the social choice rules that are compatible with the four axioms are dictatorial (in the Arrow sense). So by adding stationarity to transitivity, Pareto, and independence of irrelevant alternatives, we go from an infinite number of nondictatorial choice rules to one choice rule, which is dictatorial. And worse yet, the one remaining rule picks out the first generation as the dictator: whatever the first generation wants it gets. This result shows the strong time asymmetry built into the stationarity axiom.

Several conclusions can be drawn from the comparison of Koopmans' and Arrow's axioms. First, and most important, the comparison illustrates why the problem of intergenerational equity is deeper than that of choosing "an intergenerationally best" discount rate, $(1 - \alpha)/\alpha$. The conventional approach is to accept discounting at the level of social choice and then to attempt to choose a best "social rate of discount." But to cast the problem this way is to ignore the deeper problem, which, in the social choice context, is not what α to choose, but what axioms to choose, and what ideas of both efficiency and fairness to build into the axioms.

The deeper problem of intergenerational equity is to search for a best F; the narrower discount rate problem is to search for Fs in the form of equation (1). The problem of intergenerational equity confronts the full aggregation problem; the discount rate problem confronts a small piece of it. Further, the form of equation (1) is directly interpreted as a maximization (a maximization of discounted utilities). In this sense the problem of intergenerational equity has more to do with aggregation, and the discount rate problem more to do with maximization.

The second conclusion to be drawn is that we cannot rule out Arrow's axioms and his resulting choice rule on the basis of its inefficiency. By construction, Arrow's axioms include the Pareto axiom, and thus the resulting choice rule is intergenerationally efficient. Thus, efficiency is not a sufficient reason for defending discounting at the level of intertemporal choice. There must be some other ground.

Third, once we translate the two axiom systems, perhaps a little roughly, into operational decision criteria, they can recommend differing decisions. The divergence is most clear where there are irreversibilities and strongly asymmetric effects across time. Suppose, for example, that one is attempting to evaluate a decision about whether or not to protect an aquifer from toxic chemical contamination. Decision x is to pay large costs in the present to develop technology for chemical detoxification; decision y is to avoid the near-term costs by storing toxic chemicals in landfill sites where they leach and eventually contaminate groundwater. By assumption: (1) the carrying costs of the technological detoxification, once developed, are small; (2) the present generation, paying most attention to the costs and benefits it directly feels, prefers x to y; and (3) each future generation, also paying most attention to the costs and benefits it directly feels, prefers y to x. The profile of net benefits to each generation is shown in figure 1.

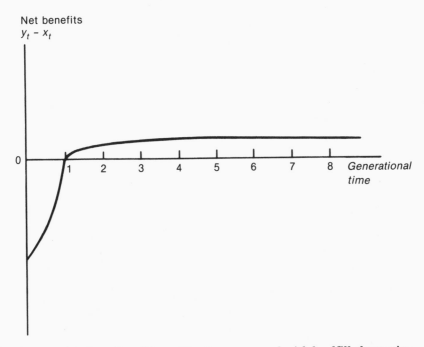

Figure 1. Net benefits of detoxification compared with landfill alternative.

For the situation in figure 1, Arrow's axioms imply that the intergenerational social choice should be y over x. However, Koopmans' axioms imply (for sufficiently low α) that the social choice should be x over y (it would be easy to develop examples where a "sufficiently low α" might be associated with traditionally recommended discount rates).

Setting aside the technical details underlying the comparison, I believe that there is some ethical appeal in the idea that the present generation should pay some current costs for later permanent benefits, but I am not saying that we should apply this idea all the time. To do so would result in the tyranny of the majority, in the intergenerational context. Just as in the case of majority voting intragenerationally, majority voting often has appeal, but not always.

Conditions of Equity as Constraints

It is convenient that Koopmans' and Arrow's axiom systems are so different in time asymmetry. With Koopmans' system strongly oriented toward the present and Arrow's toward the future, they seem to sweep out the extremes. Further, each has its intragenerational analog. Arrow's system is a little like majority rule voting. If we interpret $U(\cdot)$ in equation (1) to be consumer plus producer surplus, Koopmans' system is somewhat like traditional benefit–cost analysis and the associated acceptance of a criterion based on potential Pareto improvements.

In the intragenerational context we use both majority rule and potential Pareto improvement as decision criteria. But we use both with constraints. As a constitutional matter we may require that certain things—for example, First Amendment rights—be a part of everyone's picture, regardless of the choice over some particular project. Not necessarily as a constitutional matter but as a background condition for equity we might require that other things—such as elementary schooling—also be part of everyone's picture. In the language of social choice, such equity constraints can be written as restrictions on X.

Similarly, in the intergenerational context conditions of equity can be taken into account as restrictions on the set of feasible alternative paths $(X_1, X_2, \ldots)$. In the intergenerational context the obvious constraints have to do with the quality of the resource base that is passed on from one generation to the next. This quality defines the basic opportunities that each generation has to work with, its means of survival, and, to a large extent, its well-being. With such constraints, discounting at the level of social choice, or some more operational version of Arrow's majority rule voting, would be less problematic with respect to equity considerations.

Summary

The analogy between intragenerational equity and intergenerational equity is strong but easily obscured by the role of time in time preference and productivity of capital. Because of the role of time in the intergenerational context, discounting needs to be taken into account. By using the language of social choice it becomes clear that discounting can arise on three distinct levels: at the level of (1) social choice rules, (2) definition of the feasible set, and (3) individual preference orderings. But while it seems sensible to discount on the levels of time preference and productivity, we should not take for granted discounting at the level of social choice. The problem of intergenerational equity is deeper than simply adopting a discount rate rule of social choice and then searching for the "correct" number for the discount rate. More strongly, the comparison of Koopmans' and Arrow's axioms suggests that searching for the "correct social rate of discount" is searching for a will-o'-the-wisp.

REFERENCES

Arrow, Kenneth J., and Mordecai Kurz. 1970. *Public Investment, the Rate of Return, and Optimal Fiscal Policy* (Baltimore, Md., Johns Hopkins Press for Resources for the Future).

Dasgupta, Partha S., and Geoffrey M. Heal. 1979. *Economic Theory and Exhaustible Resources* (Cambridge, James Nisbet and Cambridge University Press).

Feldman, Allan M. 1980. *Welfare Economics and Social Choice Theory* (Boston, Nijhoff).

Ferejohn, John, and Talbot Page. 1978. "On the Foundations of Intertemporal Choice," *Journal of Agricultural Economics* (May) pp. 15–21.

Hansson, Bengt. 1976. "The Existence of Group Preference Functions," *Public Choice* vol. 38 (Winter) pp. 89–98.

Harsanyi, John. 1982. "Morality and the Theory of Rational Behavior," pp. 39–62 in A. K. Sea and B. Williams, eds., *Utilitarianism and Beyond* (Cambridge, Cambridge University Press).

Koopmans, Tjalling. 1972. "Two Papers on the Representation of Preference Orderings with Independent Components of Consumption and Representation of Preference Orderings over Time." Cowles Foundation Paper no. 366 (New Haven, Conn., Yale University Press).

Krutilla, John V. 1967. "Conservation Reconsidered," *American Economic Review* vol. 57 (September) pp. 777–786 (Resources for the Future Reprint 67).

Organisation for Economic Cooperation and Development. 1983. *Transparency for Positive Adjustment* (Paris, OECD).

Page, Talbot. 1977. *Conservation and Economic Efficiency: An Approach to Materials Policy* (Baltimore, Md., Johns Hopkins University Press for Resources for the Future).

Parfit, Derek. 1984. *Reasons and Persons*. Part four (Oxford, Clarendon Press).

Pigou, A. C. 1932. *The Economics of Welfare* (London, Macmillan).

Ramsey, Frank. 1928. "A Mathematical Theory of Saving," *Economic Journal* (December) pp. 543–559.

Solow, Robert. 1974. "The Economics of Resources or the Resources of Economics," *American Economic Review*, Papers and Proceedings, vol. LXIV, no. 2 (May), pp. 11–13.

OPTIMAL GENETIC RESOURCES IN THE CONTEXT OF ASYMMETRIC PUBLIC GOODS

Gardner M. Brown, Jr.
Joseph Swierzbinski

John Krutilla's "Conservation Reconsidered" is one of his most outstanding contributions to the field of environmental and resource economics. In that article (Krutilla, 1967; reprinted in the appendix in this volume), he argued that resource allocation decisions for unique natural resources should be based on the *ex ante* willingness to pay for the possibility of having access to the services of these resources in the future as well as on the existence and bequest values associated with them. He also set forth the idea discussed earlier by Ciriacy-Wantrup (1952) and Leopold (1949) that one source of economic value produced by species and ecosystems stems from the knowledge to be derived from them. Myers (1983) and Oldfield (1984) are among the many who have provided anecdotal evidence of the plants and animals that have enriched our lives directly and indirectly as the sources of new and better medical products, food, and industrial goods. This paper rests on John Krutilla's theme that plant and animal species are a wellspring of knowledge for society.

For thousands of years, mankind has practiced the selective breeding of animals, chosen only superior seeds for the propagation of crops, and discovered the medicinally useful constituents of plants. Discovery has grown more systematic over time, and some now label this process research and development (R&D). Clearly, we prefer a larger stock of undiscovered knowledge to a smaller one; if species are one source of knowledge, we therefore prefer to preserve species.

Yet not all species should be preserved; we should actively seek to preserve only those for which the expected net benefits of preservation are positive. Recognizing the variety of constraints to such activity, not the least of which is limited budgets, leads us to ask how many and which species should be preserved. Those species that are "selected" depend on our more or less systematic beliefs about the productivity of the phenotypes or genotypes of a species relative to alternative species. (See Brown and Goldstein [1983] for a discussion of a simple model for valuing species.) Yet there is a pronounced ethereal quality to this statement, although it has conceptual merit. We simply do not have adequate knowledge about the expected economic productivity of furbish louseworts, snail darters, and Antioch primroses, the species about which such decisions have been or are being made. Thus, current preservation decisions will necessarily be reached using qualitative analysis.

The idea of avoiding the loss of species by using the net benefit criterion holds for ecosystems or a complex aggregate of living organisms, just as it does for single species. Knowledge of that aggregate, however, is almost certainly more imprecise than that for individuals.

In an earlier paper (1983), we analyzed the desired number of species in the context of an R&D model, beginning with the observation that it is difficult to produce an optimal amount of knowledge in a market system. For example, if there are imperfect patents, a firm engaging in R&D can appropriate the benefits other firms derive, thereby creating the disincentive to invest in R&D. As the number of firms is allowed to grow, there will be no R&D in a perfectly competitive world if the results of R&D are public goods.

Yet perfect appropriability contributes to the solution of one problem by creating other problems. First, in a competitive situation in which the winner takes all, we find the well-known rushing effect as firms waste resources in order to be the first discoverers and win all (Barzel, 1968; Marglin, 1963). Second, perfect appropriability leads to monopoly power. Knowledge, which is the product of R&D and is typically small in terms of quantity, should be exchanged for the cost of transmission; it should not be sold for the cost of acquiring it. Yet, what often occurs is that too little knowledge is sold at too high a price. The consequence for species preservation is apparent. If species are a source of knowledge and the private demand for knowledge is less than the socially optimal demand because too little or too costly knowledge is produced by a competitive market structure, then the private demand for species will be less than the social demand.

There is a third disadvantage of appropriability. Let us suppose that successful R&D reduces the cost of production. (Alternatively, it may increase the quality of the product or create a new product, but these effects are conceptually quite similar.) A firm can afford to spend more on R&D if the cost spread over production is at least balanced by a sufficient cost decrease (plus an adjustment for price changes). Firms with any product market power restrict production and have less output over which to spread the cost, as shown in an earlier paper (Brown and Swierzbinski, 1983). Fourth, with any product market power, the social benefits of price decreases (consumers' surpluses) are greater than the marginal price decrease, apart from the cost of knowledge (see Spence [1984] and Brown and Swierzbinski [1983]).

Fifth, there is a strategic effect whenever there is more than one firm in an industry. When knowledge acquired by one firm spills over into other rival firms, it enhances the rivals' positions and adversely shifts the demand function of the original firm. Each firm making this strategic calculation does less R&D as a consequence. For all of these reasons, we expect any market structure to undervalue species.

As is generally true for natural resources, the source of a substantial portion of the value of any species arises as a derived demand for the species, because species contribute to the production of knowledge. Thus, our demand for knowledge can be linked to the demand and valuation of species. Too little demand for knowledge because of flaws in the market structure translates into too little demand for species. Thus, through interdependence, the value of genetic resources depends on R&D policies. The analysis that follows concentrates on R&D but considers it an economic proxy for species.

When there is too little R&D, genetic resources will be undervalued. It would be possible to compensate for inadequate demand, in part, by subsidizing firms to invest in R&D in order to achieve the socially optimal rate of such activity. We show that for a variety of product demand elasticities and other variations in industry structure, the optimal second-best subsidy of R&D expenditures is over 70 percent. For example, when the market structure is a monopoly and the product demand elasticity is constant at about 1.5, the second-best subsidy is 75 percent of the investment cost.

It is well to emphasize that these quantitative conclusions are specific to the case of constant elasticity demand functions. Indeed, when demand functions are linear, the quantitative results are different—dramatically so, in some instances. With linear demand, when there are no spillovers, the optimal subsidy is 33⅓ percent. It is more than twice as large in the nonlinear case for a demand elasticity of less than 2. The subsidy is 90 percent when the constant demand elasticity is 1.1. It can be shown that as the spillover parameter or number of firms increases, the optimal subsidy levels for the linear and nonlinear models converge. The difference between the optimal subsidy for linear demand and the constant elasticity of demand is not more than 10 percent when there are, for example, nine firms and half the knowledge is public.

The main virtue of symmetric models (those in which, for instance, all firms have the same information and in all other ways are alike) is that they are relatively simple to analyze. With them, researchers can address, as Spence (1984) so admirably does, the sort of aggregate questions we have described, such as whether or not alternative market structures produce too little research relative to a social optimum.

Yet we also need an asymmetric structure to discover the economic conditions under which a firm or firms will engage in relatively more R&D. Insofar as species "produce" the knowledge that is derived from R&D, we can draw tentative conclusions about which species are more valuable if some are relatively industry- or firm-specific.

The analysis below sets forth a model in which there are many monopolies, each producing in a separate market. Firms are not connected on the demand side. For example, there are many markets for corn seeds that are differentiated by weather, soil, pest conditions, and the price vector of substitutes. A variety developed in one region may provide the basis for developing a better variety in another region.

Introducing interdependence in the model on the demand side, as Spence has done, would create the strategic effect described in the fifth point discussed earlier (when knowledge "created" by one firm is acquired by rival firms, their position is enhanced, and the original firm's demand function is adversely affected); we know it acts as a further deterrent to privately undertaken R&D in the model being developed here. Thus, the omission of demand interdependence is qualitatively innocuous, although it may have quantitative significance. The technological interdependence among firms can easily be more important than their demand interdependence. Moreover, if the values of wild varieties of corn, wheat, and other crops are to be estimated, a multiple-market model of the sort developed in this paper will be necessary.

The model can be summarized as follows. Firms are technically interdependent on the supply side, and each firm invests in R&D. The decrease in the unit cost of production depends on a firm's level of R&D and on the level of R&D engaged in by other firms. Each firm also has its own spillover coefficient that determines the contribution of other firms' R&D activities to its success. Because the spillover effect varies across firms, the model can be viewed most generally as one in which there are heterogeneous public goods. As in earlier papers, we characterize a socially optimal level of R&D for what we term the M industries. For purposes of comparison, we reproduce the known result that, in general and for the particular case of a constant elasticity of demand and a constant marginal cost of R&D, no market structure produces an optimal amount of R&D. We then demonstrate that the greatest shortfall in R&D occurs as

1. demand elasticity decreases,
2. the number of firms sharing the results of R&D increase, and
3. the degree of spillover increases.

The last two conclusions are intuitively obvious; the first may not be as transparent. Species specific to an industry with one or more of these characteristics are leading candidates for concern because the market undervalues them. We further compute the individual R&D subsidy that maximizes social net benefits, given the profit maximization of the firms. Under most circumstances the subsidy is a very high fraction of the total cost of R&D. For example, for a demand elasticity of 1.5, it is 85 percent when the average spillover is one-third and there are three firms. The subsidy increases as the level of a firm's own contribution to others increases, and it also increases as the level of a firm's own spillover coefficient decreases.

We show further that the subsidy increases as the elasticity of demand for the product decreases. In this case, intuition would lead us to conclude that the smaller the elasticity, the smaller the output effect of a cost decrease. Thus, the cost of R&D is spread over relatively few units of output, and the subsidy is needed to reduce the fixed cost per unit of output.

Some government policies such as the investment tax credit can be viewed as a uniform subsidy across firms. In our model, we compute the optimal uniform subsidy and show that when the spillover effect (parameter) is the same for each firm, the uniform subsidy is greater than the mean of the optimal second-best individual subsidy in those cases in which the unit cost before R&D is high—that is, when R&D makes a great difference.

The Multimarket Monopoly

To begin, we describe our framework—a set of independent markets each served by one firm, but with informational interdependencies through each firm's R&D activities. In our model, there are M markets in each of which a monopolist is producing a good x_m, $m = 1, \ldots, M$, at a constant unit cost $c(R_m)$, where $c' < 0$, $c'' > 0$. The actual level of the unit cost depends on the level of research and development expenditures for all M producers. Although most of the conclusions are obtained for the more general case, in the illustrative example, it is convenient to assume that there are no diminishing returns to research expenditures. (We economize on constants by scaling research output in such a way that a firm's own research equals its own research expenditures.) Thus, the total level of R&D for each firm, R_m, is given by a linear combination of actual research expenditures r_m over all firms:

$$R_m = r_m + \theta_m \tilde{r}_m \tag{1}$$

where θ_m is the spillover parameter for the mth firm, $\tilde{r}_m$ is the sum of research expenditures for the $M - 1$ other firms, and θ_m is the marginal product to the mth firm of the efforts (expenditures) of the other firms. In later analysis, it is convenient to rewrite equation (1) as

$$R_m = r_m(1 - \theta_m) + \theta_m R \tag{1.1}$$

where

$$R = \sum^{M} r_m \quad \text{and} \quad \tilde{r}_m = \sum_{i \neq m} r_i. \tag{1.2}$$

R, therefore, is the total expenditures on R&D in the M markets.

After the optimal level of R&D is chosen in the first stage, the unit cost, $c(R_m) = \bar{c}_m$, is known for each firm, and it finds the optimal level of output that maximizes profit. Denoting the revenue of a firm by $S(x_m)$, a firm's profit, gross of research expenditures, is given by

$$\max_{x_m} \pi_m(x_m) = S(x_m) - \bar{c}_m x_m \qquad m = 1, \ldots, M, \qquad (2)$$

which it maximizes in a straightforward way.

Substituting the optimal level of output and reentering it in the profit expression yields $\pi_m(\bar{c}_m)$. Because there is an optimal level of output for any level of R&D, the gross profit expression can be written more generally as $\pi_m^0(R_m)$. The monopolist's discovery of the best level of R&D to maximize net profit in the first-stage optimization problem is then expressed as follows:

$$\text{Max } \pi_m^0(R_m) - r_m \qquad m = 1, \ldots, M, \qquad (3)$$

and solved from

$$\partial \pi_m^0 / \partial R_m = \partial \pi_m^0 / \partial r_m = 1$$

because we assume that each firm knows the levels of R&D of the other firms and takes these levels as given. Moreover, from equation (1.1), $\partial R_m / \partial r_m = 1$.

The Social Optimum

For any level of R&D there is an optimum level of output in each market. In the model, we thus suppose that the benefits of each firm, gross of research expenditures, were given by

$$G(x_m) - \bar{c}_m x_m \qquad m = 1, \ldots, M \qquad (4)$$

where $G'(x_m) = g(x_m)$ is the inverse demand curve for each product. Each firm faces a linear total (production) cost function. In this case, the optimal level of gross benefits is found by setting prices equal to marginal costs in each market:

$$g(x_m) = \bar{c}_m \qquad m = 1, \ldots, M. \qquad (5)$$

The best level of output (x_m^*) is computed from equation (5) and substituted back into equation (4) to produce a gross profit, $V_m(\bar{c}) = G(x_m^*) - \bar{c}x_m^*$. More generally, the gross benefit function for each firm for any level of R&D is denoted by $V_m(R_m)$.

The optimal level of R&D for society is determined by a process which acknowledges that research by one firm spills over to benefit other firms (profits) and their customers (consumers' surplus). Formally, the first-stage optimization problem is

$$\underset{r_i, \ldots, r_M}{\text{Max}} \sum V_m(R_m) - \sum r_m. \qquad (6)$$

The first-order condition for each r_m is

$$\partial V_m / \partial r_m + \sum \partial V_j / \partial r_m = 1, \tag{7}$$

where $\partial V_m / \partial R_m = \partial V_m / \partial r_m$ because $\partial R_m / \partial r_m = 1$.

More R&D in the aggregate occurs in the social optimum than in the monopoly. This unsurprising conclusion follows from the fact that monopolists place no weight on any consumers' surplus that a bit more R&D would produce; it also arises because monopolists, by trading output for profit, have less output over which to spread the fixed cost of R&D and because they disregard the spillover effects. The first term in equation (7) captures the first two effects whereas the second term in equation (7) accounts for the third effect. We can prove[1] that it pays each firm to have more R&D and a lower social optimum cost than its monopolist equivalent; that is, $R_m^* > R_m^N$ where, from equation (1.1), $R_m^N = (1 - \theta_m)r_m + \theta_m R$ is the monopolist's best level of R&D and r_m is its best level of expenditure, given the others' expenditures. We cannot expect that each firm with a "social conscience" would spend more than its monopolist counterpart, or $r_m^* > r_m^N$ for all m. Firms that have a large spillover coefficient and thereby benefit greatly from the R&D expenditures of others have a comparative economic advantage in letting others make socially optimal R&D expenditures.

The Social and Monopoly Multimarket Model with Structure

The examples below provide a practical application of the principles that were set out earlier in this paper and enable further conclusions to be drawn. Let us start by assuming a constant elasticity of inverse demand function for each firm:

$$g_m(x) = g_m x^{-\epsilon_m} \qquad 0 < \epsilon_m < 1, m = 1, \ldots, M \tag{8}$$

where $1/\epsilon_m$ is the price elasticity of demand in each market. From equations (4) and (8), it is easy to see that the socially optimal output (x_m^*) is

$$x_m^*(\bar{c}) = (\bar{c}_m / g_m)^{-1/\epsilon_m}$$

calculated from the first-order conditions. Substituting x_m^* in equation (4) yields

$$V_m(\bar{c}_m) = (\epsilon_m / 1 - \epsilon_m) g_m^{1/\epsilon_m} \bar{c}^{(1 - 1/\epsilon_m)}. \tag{9}$$

A similar procedure yields a gross profit for the mth monopolist of

$$\pi_m(\bar{c}_m) = (\epsilon_m / 1 - \epsilon_m) g_m^{1/\epsilon_m} (1 - \epsilon_m)^{1/\epsilon_m} \bar{c}_m^{(1 - 1/\epsilon_m)}. \tag{10}$$

Conveniently, monopoly gross profit differs from social benefits by $(1 - \epsilon_m)^{1/\epsilon_m}$, a constant that depends only on the price elasticity of demand in the market. Formally,

$$\pi_m(\bar{c})/V(\bar{c}) = (1 - \epsilon_m)^{1/\epsilon_m} = \gamma_m, \gamma_m < 1 \qquad m = 1, \ldots, M. \quad (11)$$

Because we will want to discuss a policy of subsidizing R&D expenditures, it is useful to compute social welfare levels assuming that firms will pursue profit-maximizing output levels given the subsidy. We can denote gross net benefits in this mixed case as

$$U(\bar{c}) = G(x^N) - \bar{c}x^N. \quad (12)$$

Calculating the monopolist's optimum level of output and substituting this form in equation (12) yields, for each m,

$$U_m(\bar{c}) = (\epsilon_m/1 - \epsilon_m) g_m^{1/\epsilon_m} (1 - \epsilon_m)^{1/\epsilon_m} (2 - \epsilon_m)(1 - \epsilon_m)^{-1} (\bar{c}^{1 - (1/\epsilon_m)}). \quad (13)$$

The ratio of second-best benefits to best benefits is

$$U_m(\bar{c})/V_m(\bar{c}) = \delta_m = (1 - \epsilon_m)^{1/\epsilon_m}(2 - \epsilon_m)(1 - \epsilon_m)^{-1}, \quad (14)$$

which, again, is a function only of the price elasticity of demand.

R&D Production Function

R&D is assumed to reduce the unit cost of production according to the following simple specifications:

$$c_m = c_m^0 R_m^{-a_m} \quad (15)$$

with the level of effective research, R_m, described as before in equations (1) and (1.1) and the relation between individual and total research effort summarized in equation (1.2).

The parameter a is a technological efficiency parameter characterizing the ease with which unit costs are reduced. We could introduce the supply of genetic resources (G) explicitly as $a_m(G)$, with the obvious interpretation that the greater the relevant gene pool, the easier it is to reduce unit costs to R&D and the more responsive such costs are. This method adds one more function, however, and not enough more substance to warrant the clutter.

Substituting the R&D production function (equation (15)) within the social gross benefit function (equation (9)) yields

$$V_m(R_m) = V_m^0 R_m^{\alpha_m} \qquad m = 1, \ldots, M \quad (16)$$

where

$$V_m^0 = (\epsilon_m/1 - \epsilon_m)g_m^{1/\epsilon_m} c_m^{0^{1 - 1/\epsilon_m}} \quad (16.1)$$

and

$$\alpha_m = a_m(1/\epsilon_m - 1), \qquad 0 < \alpha_m < 1. \qquad (16.2)$$

A similar substitution process using equations (15) and (10) yields the monopoly gross profit function

$$\pi_m(R_m) = \pi_m^0 R_m^{\alpha_m} \qquad (17)$$

with

$$\pi_m^0 = V_m^0(1 - \epsilon_m)^{1/\epsilon_m}; \qquad (17.1)$$

and for completeness,

$$U_m(R_m) = U_m^0 R_m^{\alpha_m} \qquad (18)$$

with

$$U_m^0 = V_m^0(1 - \epsilon_m)^{1/\epsilon_m} (2 - \epsilon_m)(1 - \epsilon_m)^{-1}. \qquad (18.1)$$

For optimum net benefits across markets, the practical problem is to

$$\max_{R_1, \ldots, R_m} \sum_{m}^{M} V_m(R_m) - R \qquad (19)$$

where individual R&D, r_m, from rewriting equation (1.1), is

$$r_m = (R_m/1 - \theta_m) - \theta_m R/1 - \theta_m \qquad (19.1)$$

and

$$R = \Sigma r_m = \Sigma (R_m/1 - \theta_m)/(1 + \Sigma \theta_m/1 - \theta_m)^{-1}. \qquad (19.2)$$

When $\theta_m = \theta$, equation (19.2) has a simple expression:

$$R = \Sigma R_m/[1 + (m - 1)\theta]. \qquad (19.3)$$

Because the firm chooses individual R&D expenditures, r_m, it is natural to maximize in terms of these individual choice variables. However, solving in terms of R_m is more straightforward because the r_ms can be computed from the R_ms.

The socially best level of effective R&D, R_m^*, (that is, when each firm does some), which can be found from maximizing social net benefits in equation (19), using the expression for total individual research expenditures ($\Sigma r_m = R$) in equation (19.3), is as follows:[2]

$$R_m^* = \{[1 + (M - 1)\theta]\alpha_m V_m^0\}^{1/1 - \alpha_m}. \qquad (20)$$

For economy, $\theta_m = \theta$ until it is important to emphasize distinctions among firms.

The profit-maximizing level of effective R&D, R_m^N, with each firm engaging in R&D, is found from maximizing the monopoly gross profit (equation (17)) minus the cost of R&D, or

$$\max_{r_m} \pi_m^0 R_m^{\alpha_m} - r_m, \qquad m = 1, \ldots, M. \tag{21}$$

We obtain

$$R_m^N = (\alpha_m \pi_m^\theta)^{1/1 - \alpha_m} \tag{22}$$

From the exact relationship between profit and social benefits in equation (11), equation (22) can be rewritten as

$$R_m^N = (\alpha_m V_m^0 \gamma_m)^{1/1 - \alpha_m} \tag{23}$$

so that effective R&D for each firm is greater in the socially optimum situation. That is,

$$R_m^*/R_m^N = \{[1 + (M - 1)\theta]/\gamma_m\}^{1/1 - \alpha_m}, \tag{24}$$

where γ_m and $\alpha_m < 1$. It is easy to show that

$$R_m^*/R_m^N = r_m^*/r_m^N.$$

There are two key sources of difference between optimal and monopoly-effective R&D. Whenever there is no competition (apart from perfect discrimination or its equivalent), realized output for a given unit cost falls short of competitive output because marginal benefit is greater than marginal profit. This wedge, which is captured by γ_m (see equation (11)), distorts the choice of R&D effort. Second, spillovers benefiting others, which are disregarded by monopolists, are captured by $(M - 1)\theta$. If there were only one firm ($M = 1$), the numerator would be the same for the social planner as for the monopolist.

Socially effective R&D increases relative to the monopoly level as the spillover benefits increase—as θ or θ_m increases—and as the number of firms (M) sharing the public goods increases. The discrepancy between social and private levels of R&D grows as the elasticity of demand ($1/\epsilon_m$) decreases (as ϵ increases) (figure 1).[3] The less elastic the demand, the less output responds to a price decrease occasioned by cost-decreasing R&D. Finally, the greater the elasticity of the unit cost decrease as a result of R&D (the greater the value of a in equation (15)), the greater the value of R_m^*/R_m^N because a and α vary directly (see equation (16.2)).

The relationship between private research and socially optimal research levels is illustrated in table 1 using the assumption that the technological efficiency parameter is 0.1.[4] Inverting the ratio, putting optimal research in the denominator, usefully confines comparisons to fractions less than 1. Performance under the various private regimes is not particularly stellar. Private

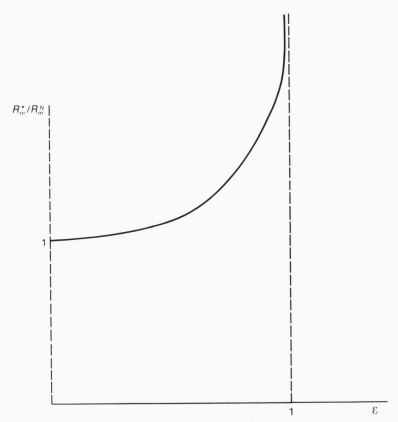

Figure 1. The divergence between optimal and private R&D.

R&D never exceeds 21 percent of the socially optimum level in these cases; more typically, it is under 10 percent. It is interesting to compare welfare values, the sum of producers' and consumers' surplus in the social (W^*) cases. The comparison is a relative one, W^N/W^*, with each computed at its respective optimal levels of output and R&D. Table 2 illustrates the comparison for various combinations of the spillover coefficient, demand elasticities, and number of firms. Relative welfare levels do not produce as striking a contrast as the relative levels of R&D produce. Many firms and high demand elasticities make private welfare relatively low. For more reasonable demand elasticities, private welfare is over 50 percent of the optimal levels, rising to around 70 percent or more for demand elasticities under 2 and for modest spillover coefficients.

A Subsidy to Achieve a Second-Best Optimum

In view of the relatively poor performance achieved by private firms when some fraction of private R&D is a public good, is there an effective policy that

Table 1. Private Relative to Optimal Research Levels per Firm (r^n/r^*)
(Demand elasticity = 1/ϵ)

No. of firms (M)	Spillover θ = .1				Spillover θ = .5				Spillover θ = .8			
	1.1	1.5	2.0	5.0	1.1	1.5	2.0	5.0	1.1	1.5	2.0	5.0
1	0.07	0.18	0.21	0.16	0.07	0.18	0.21	0.16	0.07	0.18	0.21	0.16
2	0.06	0.16	0.19	0.13	0.05	0.12	0.14	0.08	0.04	0.10	0.11	0.06
5	0.05	0.12	0.15	0.09	0.02	0.06	0.06	0.02	0.02	0.04	0.04	0.01
10	0.04	0.09	0.11	0.05	0.01	0.03	0.03	0.01	0.01	0.02	0.02	0.00

Table 2. Private Relative to Optimal Welfare (W^N/W^*)
(Demand elasticity = 1/ϵ)

No. of firms (M)	Spillover θ = .1				Spillover θ = .5				Spillover θ = .8			
	1.1	1.5	2.0	5.0	1.1	1.5	2.0	5.0	1.1	1.5	2.0	5.0
1	0.84	0.73	0.69	0.48	0.84	0.73	0.69	0.48	0.84	0.73	0.69	0.48
2	0.84	0.73	0.69	0.46	0.84	0.72	0.67	0.39	0.84	0.72	0.66	0.36
5	0.84	0.72	0.67	0.41	0.83	0.70	0.63	0.26	0.83	0.69	0.60	0.21
10	0.84	0.71	0.65	0.35	0.83	0.68	0.59	0.18	0.83	0.66	0.56	0.15

could improve private performance? Private firms could be encouraged to engage in more R&D by being paid a fraction, s_m, of their R&D expenditures as a subsidy. We can regard s_m as an R&D investment tax credit. Instead of the firm's maximizing profit as expressed in equation (21), its new problem is

$$\max_{r_m} \pi_m^0 R_m^{\alpha_m} - (1 - s_m)r_m; \tag{25}$$

because $\partial R_m/\partial r_m = 1$, the first-order condition for each firm is

$$\alpha_m \pi_m^0 R_m^{\alpha_m - 1} = 1 - s_m. \tag{26}$$

Earlier, in equation (18), the level of gross benefits for any choice of R&D by a firm was computed to be $U_m^0 R^{\alpha_m}$. The second-best level of net benefits is therefore

$$W(R_1, \ldots, R_M) = \Sigma U_m^0 R_m^{\alpha_m} - \Sigma r_m,$$

rewritten as

$$\tilde{W}(R_1, \ldots, R_m) = \Sigma U_m^0 R_m^{\alpha_m} - \Sigma \phi_m(R_m), \tag{27}$$

where from equation (19.2),

$$\Sigma r_m = \Sigma(R_m/1 - \theta_m)/(1 + \Sigma \theta_m/(1 - \theta_m)). \tag{28}$$

The second-best level of R_m is found by solving

$$\alpha_m U_m^0 R_m^{\alpha_m - 1} = \phi_m'(R_m). \tag{29}$$

From earlier calculations in equations (11) and (14),

$$U_m^0 = \pi_m^0 \delta_m / \gamma_m,$$

which, after substitution into equation (29) and revising, yields

$$\alpha_m \pi_m^0 R_m^{\alpha_m - 1} = (\gamma_m/\delta_m)\phi_m'(R_m). \tag{30}$$

All we must do now is to pick the best subsidy level, $1 - s_m$, to induce the private firm to switch its marginal profit-maximizing rule for choosing R_m (or r_m) in equation (26) to the one that solves equation (30) and achieves the second-best social optimum. It is easy to see from equations (26), (28), and (30) that

$$1 - s_m = \gamma_m/\delta_m(1 - \theta_m)(1 + \Sigma \theta_m/(1 - \theta_m)). \tag{31}$$

When spillovers are symmetric, $\theta_m = \theta$, then

$$1 - s_m = \gamma_m/\delta_m[1 + (M - 1)\theta], \tag{32}$$

a more familiar expression (because $\gamma_m/\delta_m < 1, s_m > 0$).

Not surprisingly, the subsidy to a firm decreases as its spillover parameter increases. Other firms are relatively more productive. However, in the symmetric case described by equation (32), either a larger spillover or a greater number of firms make R&D expenditures socially more productive; consequently, s_m should increase. Finally, it is easy to show from the definition of γ_m/δ_m that the subsidy increases as the elasticity of demand decreases. These results are illustrated in table 3. Subsidies are a large fraction of the total research expenditure. When there is monopoly, the subsidy varies from 56 to 90 percent depending on the elasticity of demand. The same values hold for each elasticity, regardless of the number of firms, when the spillover is 0. When θ is low (0.1), the subsidy is 96 percent when there are 10 firms benefiting from the spillover and the demand elasticity is low (1.1); the subsidy is 76 percent when demand elasticity is 77 percent. When the spillover is large (0.8), the comparable subsidy levels are 99 and 95 percent.

To see how effective an optimal subsidy can be in inducing research levels to approach the socially desired levels approximately, let us suppose there are M identical firms and then compare the level of subsidized research (r^s) with the socially optimal level of research (r^*). From equations (30), (31), (20), (1.1), and (1.2), it can be shown that

$$r^s/r^* = (U_0/V_0)^{1/1-\alpha} = [(1 - \epsilon)^{1/\epsilon}(2 - \epsilon)/(1 - \epsilon)]^{1/1-\alpha},$$

which is independent of the number of firms and the magnitude of the spillover parameter. (Also, $R^s/R^N = r^s/r^N$.)

Because the ratio of welfare with a subsidy (W^s) and socially optimal welfare (W^*) is

$$W^s/W^* = \{U_0[r^s[1 + (m - 1)\theta]]^\alpha - r^s\}/\{V_0[r^*[1 + (m - 1)\theta]]^\alpha - r^*\},$$

it is easy to show that

$$W^s/W^* = r^s/r^*.$$

Table 4 illustrates welfare and research levels for the case of an optimal subsidy level relative to the socially optimal levels. We can draw three important conclusions from table 4. First, under reasonable ranges for the parameters, the level of research responds dramatically to an optimal second-best subsidy with a concomitant payoff in welfare. For a demand elasticity of 1.5 and an innovative efficiency of 0.1, research levels increase by more than a factor of 4. Compare tables 4 and 1 in which relative research, r^n/r^*, is never greater than 17.6 percent, and is as low as 2 percent without a subsidy, for the number of firms varying between 1 and 10 when $1/\epsilon = 1.5$. Yet r^n/r^* is 76 percent with a subsidy. As demand becomes more elastic, the welfare effect of a subsidy diminishes, but the diminution is substantial only at high levels of market demand elasticities. At the limit, of course, the subsidy is completely ineffective. There is no consumers' surplus to convert into monopoly profit when a

Table 3. Optimal Subsidy, in Percent
(Demand elasticity = $1/\epsilon$)

No. of firms (M)	Spillover θ = 0				Spillover θ = .1				Spillover θ = .5				Spillover θ = .8			
	1.1	1.5	2.0	5.0	1.1	1.5	2.0	5.0	1.1	1.5	2.0	5.0	1.1	1.5	2.0	5.0
1	92	75	67	56	92	75	67	56	92	75	67	56	92	75	67	56
2	92	75	67	56	92	77	70	60	94	73	78	70	95	86	81	75
5	92	75	67	56	94	62	76	68	97	92	89	85	98	94	92	89
10	92	75	67	56	96	87	72	77	98	95	94	92	99	97	96	95

Table 4. Second-Best Optimal Subsidy Welfare and Research Levels Relative to Socially Optimal Performance (w^s/w^*, R^s/R^*)

Technological coefficient A	Demand elasticity			
	1.1	1.5	2.0	5.0
.05	.86	.76	.73	.69
.1	.86	.76	.73	.61
.2	.86	.75	.70	.22

monopolist faces a completely elastic demand function. Thus, there is, in the limit, no reward for innovating to bring production costs down.

Second, although research levels respond dramatically to a subsidy, welfare does not, as can be seen by comparing tables 2 and 4. For low elasticities of demand and few firms, there is hardly any improvement in welfare. Illustratively, relative welfare with an optimal second-best subsidy improves to 73 percent of optimal, from 67 percent for $1/\epsilon = 2.0$ and five markets, despite the fact that research levels have increased by nearly a factor of 5. (This same result can be discovered in Spence (1984) by comparing the appropriate tables, although there is no reference to the curiosity in the text.) We are currently exploring the reasons for this puzzle. Part of the explanation may be that the lower the elasticity, the lower the output. At lower levels of output, R&D reduces unit costs, but the gross gain in welfare is almost wiped out by (in effect) spreading the cost of R&D over relatively fewer units of output. To put it another way, what is gained in reduced cost is lost to a large extent in R&D expenses when output is low. When the number of firms and the elasticity of demand grows, performance improves markedly, for example, 50 percent. However, the overall performance is fairly mediocre under 25 percent of the optimum for $1/\epsilon = 5$ and $M = 10$.

Finally, the productivity of a subsidy is not much affected by the ease with which cost reductions are achieved.

The parameter a is the elasticity of unit cost reductions with respect to research. Doubling or tripling its level around the magnitude of empirical estimates (0.1) has little effect on performance.

Nash Duopoly in R&D Expenditures

In this model, it is analytically convenient to assume that all firms are the same, but such an assumption may not be descriptively accurate. When each of the M firms is allowed to differ in demand elasticity, in spillover parameters, or in innovative efficiency, the ability to reach general conclusions disappears. Some feeling for the consequences of differences among firms can be obtained by reducing the number of firms to two, imagining, perhaps, that each firm is an appropriately aggregated subset of other firms.

Recalling the definition of R_m from equation (1.1), equation (22) can be written as

$$r_1 = \bar{r}_1 - \theta_1 r_2, \tag{33}$$

where

$$\bar{r}_1 = \alpha_1 \pi_1^{0^{1/1 - \alpha_1}},$$

the optimal level of firm 1's R&D if it enjoys no spillovers, that is, $\theta_1 = 0$. Similarly, the reaction function for firm 2 is

$$r_2 = \bar{r}_2 - \theta_2 r_1. \tag{34}$$

Illustrative reaction functions and the Nash equilibrium are pictured in figure 2a. Yet other characterizations are possible. Firm 1's spillover could be small, for example, $\theta_1 = 0$; thus, $r_1 = \bar{r}_1$, and firm 2 does no research, the solution depicted by figure 2c. Alternatively, if firm 1's beneficial spillover (θ_1) is large, it might be optimal for it to do no research ($r_1 = 0$), so that $r_2^N = \bar{r}_2$ and the reaction functions do not intersect, as shown in figure 2b. This is an interesting case because it is easy to show reasonable conditions under which it would pay some profit-maximizing firms *not* to do research; yet these same firms *ought* to do research (and the others should not) in the socially optimal case.

The contradiction between the social and private optimum is readily apparent in the context of two firms. In the private, noncooperative case, firm 1 does no research if, from equation (22) and the definition of R_m,

$$\theta_1 \geq (\alpha_1 \pi_1^0)^{1/1 - \alpha_1}/r_2^N = (\alpha_1 \pi_1^0)^{1/1 - \alpha_1}/(\alpha_2 \pi_2^0)^{1/1 - \alpha_2}. \tag{33.1}$$

The condition for firm 2 doing no research for a social optimum, which is derived from note 2 and the definition of R_m, is

$$\theta_2 \geq [\alpha_2 V_2^0 (1 - \theta_1 \theta_2)(1 - \theta_1)^{-1}]^{1/1 - \alpha_2}/$$
$$[\alpha_1 V_1^0 (1 - \theta_1 \theta_2)(1 - \theta_2)^{-1}]^{1/1 - \alpha_1}. \tag{34.1}$$

Because α_m is a combination of the R&D technology parameter (a_m) and the elasticity of demand (ϵ_m^{-1}) while V_m^0 and π_m^0 depend on the elasticity of demand and the firm's size (c_m and g_m), ten parameters, including the spillover parameters, contribute to the contradiction, making it difficult to focus sharply on the driving forces. It is true that the larger firm 1's spillover parameter, the more easily the conditions are met. More generally, each firm scrutinizes the benefits and costs of its R&D and output decisions in a different manner than its socially managed counterpart. A firm neglects consumers' surpluses in its

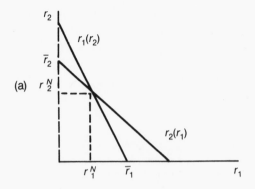

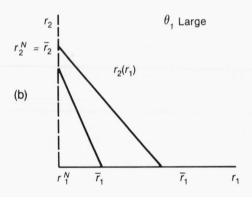

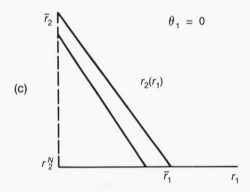

Figure 2(a–c). Reaction functions of Nash duopolists in R&D expenditures.

own and in the other market. It also neglects the cost decreases in other markets that its R&D investment occasions through the spillover.

The most straightforward way to simplify the conditions is to introduce differences in the relative scale of firms (g_m), letting $\alpha_1 = \alpha_2 = \alpha$, $\epsilon_1 = \epsilon_2 = \epsilon$, and $c_m = 1$. Then, from equations (10), (13), (33.1), and (34.1), firm 1 does no research in the private setting if

$$\theta_1 \geq (g_1/g_2)^{1/\epsilon(1-\alpha)}, \tag{35}$$

and firm 2 does no research in the socially optimum situation if

$$\theta_2 \geq [(1 - \theta_2)/(1 - \theta_1)\,(g_1/g_2)^{1/\epsilon}]^{1/1-\alpha}. \tag{36}$$

It is apparent from equation (35) that the greater the natural scale of firm 2, the lower firm 1's spillover (θ_1) can be and yet still warrant no research. This condition makes sense because the scale of production is synonymous with spreading the fixed cost of research and encouraging research by firm 2 in this instance. For the same reason, as firm 2's scale increases, the higher its spillover parameter (θ_2) must be before it forgoes the gains from its own research and relies on firm 1's research activities.

Another way to simplify the conditions is to let $\alpha_1 = \alpha_2$ and let $g_m = c_m = 1$. Then, from equations (10), (13), (33.1), and (34.1), firm 1 does no research in the private setting if

$$\theta_1 \geq [\epsilon_1(1 - \epsilon_1)^{1 - \epsilon_1/\epsilon_1}/\epsilon_2(1 - \epsilon_2)^{1 - \epsilon_2/\epsilon_2}]^{1/1-\alpha},$$

and firm 2 does no research in the public setting if

$$\theta_2 \geq [\epsilon_1(1 - \epsilon_2)(1 - \theta_2)/\epsilon_2(1 - \epsilon_1)(1 - \theta_1)]^{1/1-\alpha}.$$

We can see from these conditions that if the spillover parameters are the same $(\theta_1 = \theta_2)$, the wrong firms will do research in the private case if the demand elasticities are chosen in an appropriate manner. This, of course, is not surprising because the demand elasticity determines consumers' surpluses, which are regarded differently by firms and society. However, we cannot easily say which demand elasticity must be higher. Because $\alpha = a[(1/\epsilon) - 1]$, assuming a common α, as we did, implies that the firm with the higher demand elasticity $(1/\epsilon)$ must be assigned a smaller technology parameter (a), which reduces the marginal product of R&D investment. Thus, the admixture of changes in demand and supply parameters blocks easy intuition.

When firms are distinctive and are technologically interdependent, their aggregate performance, not surprisingly, depends on the distinctions among them, making it impossible to reach general quantitative conclusions. In the cases below (table 5), we have tried to choose some variation to give readers a feeling for the results. The reference point is case 1 in which firm 1 has a relatively low elasticity of demand of 1.25 (it must be greater than 1), and

Table 5. Relative Performance and Qualitative R&D Activity of Private, Subsidized, and Optimally Managed Duopolists (percentage)

Type of Duopolist	Case 1[a]				Case 2[b]				Case 3[c]				Case 4[d]			
	(a) $\theta=.8$ $\theta_2=.1$	(b) $\theta=.8$ $\theta_2=.1$	(c) $\theta=.1$ $\theta_2=.8$	(d) $\theta=.1$ $\theta_2=.8$	(a) $\theta=.8$ $\theta_2=.1$	(b) $\theta=.1$ $\theta_2=.8$	(c) $\theta=.1$ $\theta_2=.8$	(d) $\theta=.8$ $\theta_2=.1$	(a) $\theta=.8$ $\theta_2=.1$	(b) $\theta=.1$ $\theta_2=.1$	(c) $\theta=.1$ $\theta_2=.8$	(d) $\theta=.8$ $\theta_2=.8$	(a) $\theta=.8$	(b) $\theta=.1$ $\theta_2=.1$	(c) $\theta_2=.1$	(d) $\theta_2=.8$
$\Sigma r_m^N/\Sigma r_m^*$	6	15	4	3	13	17	5	4	11	14	4	4	10	19	3	21
$\Sigma r_m^S/\Sigma r_m^* = W^S/W^*$	79	79	79	79	79	79	79	79	78	78	78	78	69	69	69	69
W^N/W^*	13	13	13	14	14	14	14	14	9	9	9	9	7	17	11	17
W^S/W^*	79	79	79	79	79	79	79	79	78	78	78	78	69	69	69	69

Qualitative Characterization of R&D Activity[e] for Duopolists[f]

	Case 1				Case 2				Case 3				Case 4			
	(a)	(b)	(c)	(d)	(a)	(b)	(c)	(d)	(a)	(b)	(c)	(d)	(a)	(b)	(c)	(d)
Private	1	Both	Both	Both	2	Both	2	Both	Both	Both	Both	Both	2	2	2	2
Optimal	Both	Both	2	1	1	Both	2	1	Both	Both	2	1	2	2	2	2

[a] $a = 0.1$; $g_m = c_m = 1$; $\epsilon_1^{-1} = 1.25$; $\epsilon_2^{-1} = 3$; $\theta_1 = .8$; and $\theta_1 = .1$.
[b] Same as case 1 but innovation more inelastic ($a = .05$).
[c] Same as case 1 but closer elasticities ($\epsilon_1^{-1} = 1.25$; $\epsilon_2^{-1} = 1.67$).
[d] Same as case 1 but larger scale ($c_m = 1$; $g_m = 100$).
[e] 1 = firm 1; 2 = firm 2.
[f] Designates which firm undertakes R&D activity.

firm 2 has a relatively large elasticity of demand of 3.0. Case 1(a) has a large common spillover parameter ($\theta = .8$), as do the other (a) subcases. Case 1(b) has a small spillover parameter ($\theta = .1$), as do the other (b) subcases. In case 1(c), firm 1 has a large spillover parameter whereas firm 2's is small; in case 1(d) the size of the spillover parameters is reversed. Case 2 is introduced to show the result of halving the elasticity of innovation (a) from 0.1 to 0.05. Case 3 indicates how the results are affected if the demand elasticities are brought closer together. Finally, case 4 portrays two firms with larger markets than in case 1 as captured by a larger constant term (g_m) in the inverse demand function.

It is clear from table 5 that private research is never a large fraction of the socially optimal level, exceeding 20 percent in only one instance (case 4[d]). Introducing a second-best subsidy (which restores the incentive to do R&D but does not address distortions from output restrictions) greatly improves performance, as can be seen if we compare row 2 to row 1 of the table. Subsidized research levels rise to 69 or more percent of the optimum level (row 2), and there is a dramatic improvement in society's net benefits. Net benefits from R&D without a subsidy are generally not more than about 14 percent of the possible (row 3); the subsidy raises the ratio to about 70 percent or more. Subsidies improve performance the most, relatively speaking, when elasticities are relatively small (case 3).

The lower portion of table 5 summarizes which firms do research in the sixteen cases. First, when the spillover parameter is small (the [b] cases), both firms do research when scale is not significant (cases 1, 2, and 3). Second, the greatest qualitative discrepancy between private R&D and the optimal distribution of R&D arises when the spillover parameter is large. In seven out of nine cases of small markets (cases 1, 2, and 3), when $\theta_m = .8$ (cases [a], [c], and [d]), the firm(s) doing R&D under optimal conditions differ from those doing R&D in the nonsubsidized duopoly setting. In case 2(a), in fact, firm 1 does R&D optimally, but firm 2 is the only firm doing R&D in the duopoly setting. Scale, however, combined with firm 2's relatively large demand elasticity, dominates the differences in spillover in case 4.

A Uniform Subsidy Policy

It is often the case that general legislation does not recognize differences among individual firms. A uniform R&D investment tax credit of s percent of investment is an example. Since 1981 the uniform subsidy in the United States has been a 25 percent tax credit (Mansfield, 1985).[5] What would a uniform subsidy look like, and how does it compare with the second-best subsidy tailored to each firm/industry discussed earlier in this chapter? Instead of responding to s_m, the firm responds to s and chooses the best level of R&D

investment according to the marginal conditions of equation (26), which is rewritten here as

$$R^u(s) = (\alpha \pi_m^0/(1 - s))^{1/1 - \alpha} \qquad (37)$$

where the superscript u denotes uniform. From equation (37),

$$\partial R^u(s)/\partial s = R^u(s)/(1 - s)(1 - \alpha). \qquad (38)$$

Subsequent analysis is conducted with symmetric spillovers, $\theta_m = \theta$ and $\alpha_m = \alpha$. The problem for those setting the subsidy is to find the uniform subsidy that maximizes social net benefits, given that each firm reacts to s according to equation (37). For the planner, then, the problem is

$$\max W^u - \Sigma[U_m^0 R_m^\alpha(s) - R_m(s)/1 + (M - 1)\theta], \qquad (39)$$

a natural adaptation of equation (27).

After tedious calculations using $\pi_m^0 \delta_m/\gamma_m = U_m^0$ from equations (11), (14), and (38) in the first-order conditions, we obtain

$$(1 - s^u) = \Sigma \pi_m^{0(1/1 - \alpha)}/[1 + (M - 1)\theta]\Sigma(\delta_m/\gamma_m)\pi_m^{0(1/1 - \alpha)},$$

rewritten as

$$1 - s^u = ([1 + (M - 1)\theta]\Sigma \delta_m/\gamma_m * P_m/\Sigma P_m)^{-1} \qquad (39)$$

where $P_m = \pi_m^{0(1/1 - \alpha)}$. $P_m/\Sigma P_m$ can be thought of as probabilities or weights that sum to 1. Thus, if $\bar{u}$ is the mean of δ_m/γ_m, $\Sigma(\delta_m/\gamma_m)P_m/\Sigma P_m = \omega\bar{u}$ is the (profit) weighted mean of δ_m/γ_m, so

$$1 - s^u = ([1 + (M - 1)\theta]\omega\bar{u})^{-1},$$

which can be compared with the mean of the individual subsidy computed from equation (32),

$$1 - \bar{s} = ([1 + (M - 1)\theta]\bar{u})^{-1}.$$

Because firms with relatively large profits have larger weights and the individual subsidy varies with profit, the uniform subsidy is greater than the average subsidy, and the empirical illustrations bear out this reasoning.

It is instructive to compare performance under a uniform subsidy policy with performance using optimal subsidies. As a general principle, we would expect to observe the greatest divergence when the parameters determining the subsidy are increasingly disparate among the firms. To simplify the expressions, we have assumed there are no differences in α and θ, leaving only variations in ϵ for comparisons. However, variations in the magnitude of θ can also be compared.[6]

Table 6 summarizes three comparisons between the second-best optimal subsidy and a uniform subsidy. Not surprisingly, the greatest divergence in

Table 6. Comparing Performance Between a Best Uniform and a Best Variable Subsidy

	Variation in demand elasticities		
	Larger[a]	Moderate[b]	
	Low spillover[c] (1)	Low spillover[c] (2)	High spillover[d] (3)
Average variable subsidy ($\bar{S}$)	.75	.80	.88
Uniform subsidy (S^u)	.83	.82	.89
Effective research with $\bar{S}$	.13	.15	.26
Effective research with S^u	.18	.21	.41
Welfare with $\bar{S}$	3.21	3.8	3.9
Welfare with S^u	3.06	3.7	3.8

[a] $\epsilon_1 = 1.25; \epsilon_2 = 3$
[b] $\epsilon_1 = 1.25; \epsilon_2 = 1.6$
[c] $\theta = .1$
[d] $\theta = .8$

welfare occurs when the elasticities between the two firms are most different (compare columns 2 and 1). We suspect the absolute difference is small because of the choice of scaling factors. Performance measured in terms of welfare is not very much altered by changing the magnitude of the spillover parameter (compare columns 2 and 3).

The uniform subsidy of 25 percent offered by tax relief in the United States is small compared to the subsidies that emerge from the model and the functions specified in this paper. It is easy to show that the optimal second-best subsidy should never be less than 0.5, a situation that occurs when, for example, $\theta = 0$ or there is monopoly. In these cases the uniform subsidy is over 0.8, compared with the legislated 0.25. Both subsidy rates increase as θ or M increase.

Conclusion

If the rate of extinction of species is too high, it may be occurring because there are flaws in the market structure. Identifying these flaws is necessary before remedial policies can be designed. The more precisely the flaws are described, the more efficiently policies for reducing extinction can be developed and defended. It is common knowledge that genetic resources are directly and indirectly the source of knowledge discovered through the R&D process. In this paper, we characterize the public goods nature of R&D by introducing a parameter that accounts for the spillover of research results to a firm from other firms' R&D activities. The leakage of some value of the R&D product from the firm paying for the genetic resources input creates inefficiency and social waste.

The externality veils actual demand for the product, which is then reflected in an inward shift of the demand for the genetic resources input, thus causing its systematic undervaluation.

The spillover externality is completely different from the common property aspect of the genetic resources themselves and would arise even if all genetic resources were privately owned and properly managed. The fact that they are not suggests that the stock of genetic resources is suboptimal. This notion is captured implicitly in the model by the technological efficiency parameter (a), which characterizes the ease with which R&D reduces unit cost or increases product quality. A lower stock of genetic resources makes it harder to innovate. John Krutilla and others have discussed these social costs—costs that are caused by the failure to manage properly the common property externality of genetic resources. This paper explores the negative feedback effect of the common property aspect of R&D. Because of our assumed linkage between the genetic stock as reflected in the species on this planet, we have also provided a connection between this open-access effect and the valuation of genetic resources.

The basic structure of this paper resembles recent work by Dasgupta and Stiglitz (1980) and by Spence (1984) inasmuch as multiple firms reach a Nash–Cournot equilibrium in R&D expenditures. Furthermore, the spillover of R&D and the research subsidy as a policy analogue are features common to Spence's paper and ours. Yet the models of these other authors have a single-market monopolistic competition structure whereas ours is a multimarket structure with firms linked technologically, on the supply side, by the R&D spillover. The extension to multiple markets is important because substantial R&D is germane for more than one market. For example, R&D financed by defense expenditures has, in turn, led to new forms of commercial aircraft, new materials, better freeze-dried foods, and other innovative products. New packaging ideas have multiple-market applications of special interest to us. Certain wild seed varieties have been used to develop an array of seeds, each variety of which is specially designed for a particular geographic market. Indeed, the multimarket use of a developed idea seems to be the rule and not the exception.

Our initial treatment of firms as identical further resembles Spence (1984) and Dasgupta and Stiglitz (1980) but we come closer to reality by studying firms that differ in three ways: (1) the amount of free R&D they receive, (2) demand elasticity, and (3) technological efficiency. Thus, the optimal subsidy for each firm varies.

Firms that are monopolists in our model perform too little R&D as compared with socially optimum levels because they give no weight to the benefits that spillovers confer on other firms and because they pursue profit and not consumers' (and producers') surpluses. By restricting output to capture profit, there is less volume over which to spread the fixed cost of R&D. The divergence between actual and optimal R&D naturally increases as the spillover factor and number of firms (markets) increase. The divergence also increases as the elasticity of demand decreases because such a situation is the same as a smaller market in our model. It is also reasonable that a subsidy designed to price the externality should vary directly with the size of the externality. It increases with the number of markets and the size of spillover effect and decreases with the demand elasticity. The subsidy, however, is a second-best subsidy: it encourages the firms to do a socially optimal amount of R&D but only in the event they produce monopoly levels of output at that level of R&D.

The subsidy levels seem quite high—more than 70 percent of the cost of R&D—for what we believe are plausible values of the parameters: an elasticity of demand of about 1.5 and a spillover parameter of 10 to 50 percent. If our beliefs are correct—the reader being cautioned that there is not a great deal of specific research to guide intuition—then the species we should seriously consider saving are those that are contributing or that might contribute to the creation of knowledge in the industries whose characteristics warrant high subsidies. Those subsidies, however, may not exist. (We use the word "may" in this case because a study of the commercial seed industry would probably conclude that the nation's seed banks are an important form of subsidy. Whether seed banks are enough of a subsidy, however, is an empirical question that begs study.)

When asymmetry among firms is introduced, we can see the system in greater detail, but the general conclusions just discussed continue to hold. Private research is low relative to optimal research; it rarely exceeds 20 percent, and subsidy levels are high (75 percent or greater for the parameters chosen). Both of these "stylized facts" indicate the dramatic undervaluation of genetic resources specific to R&D in industries described by the broad range of parameters we have studied. We discovered that in most (seven out of nine) of the illustrative cases studied, the firm(s) that, in an optimal setting, should conduct research are not the firms that actually do it in a private, unsubsidized setting. This could be an artifact of our choices of parameter values, but we do not think so. The result suggests that, if species are specific to particular firms or industries, society would save a different array of species than that saved by private industry. This, of course, is a speculation; we are aware of no research results that can refute or corroborate this proposition.

Finally, we have compared an optimal second-best uniform subsidy for all with an individual subsidy. The uniform subsidy is larger and produces more research, but it concomitantly produces less welfare. There is, to a first approximation, a uniform subsidy for R&D in the United States in the form of a tax credit, but it is perhaps three or more times smaller than the level suggested by our examples.

John Krutilla did not and would not look at genetic resources from the economic perspective we have taken. Nevertheless, he addressed resource conservation issues in a way that was dramatically different from the approach used by Ciriacy-Wantrup, our mutual intellectual forebear. Yet both contributed toward an understanding of the importance of preserving genetic resources. This paper was written in the same spirit.

NOTES

1. From the definitions of $\pi_m^0(R_m)$, $V_m(R_m)$, and R_m, $V_m(R_m) - \pi_m^0(R_m)$ is concave; thus, $R_m^* > R_m^N$ for all m, where R_m^N is the optimal R_m for the monopolist. For some but not necessarily all firms, $r_m^* > r_m^N$. However, $R^* > R^N$ because from equation (1.1)

$$r_m = R_m/(1 - \theta_m) - \theta_m R/(1 - \theta_m)$$

$$R = \Sigma r_m = [\Sigma R_m/(1 - \theta_m)]/1 + \Sigma \theta_m/(1 - \theta_m)$$

and we have demonstrated that $R_m^* > R_m^N$ for all m.

2. The first-order condition, in general, is

$$\alpha_m V_m^0 R_m^{\alpha_m - 1} = 1/(1 - \theta_m)(1 + \Sigma \theta_m/(1 - \theta_m)).$$

3. Equation (24) may be rewritten, after dropping subscripts on the right-hand side, as

$$R_m^*/R_m^N = \left[\frac{1 + (M - 1)\theta}{\gamma}\right]^{\frac{1}{1 - \alpha}} = \left[\frac{1 + (M - 1)\theta}{\gamma}\right]^{\frac{\epsilon}{\epsilon - a(1 - \epsilon)}}$$

because $1 > \alpha = a\left(\dfrac{1}{\epsilon} - 1\right)$. The $\lim\limits_{\epsilon \to 0}\left(\dfrac{\epsilon}{\epsilon - a(1 - \epsilon)}\right) = 0$ because

$\lim\limits_{\epsilon \to 0} a\,\epsilon = 0$. To see this, note that $\alpha = a\left(\dfrac{1}{\epsilon} - 1\right)$ can be rewritten as

$\dfrac{\alpha\,\epsilon^2}{1 - \epsilon} = a\,\epsilon$. Clearly, the right-hand side approaches 0 as $\epsilon \to 0$, so the

left-hand side also must approach 0. Additionally, rewrite $\gamma^{\frac{-\epsilon}{\epsilon - a(1 - \epsilon)}}$ as

$(1 - \epsilon)^{\frac{-1}{\epsilon - a(1 - \epsilon)}}$ using the definition of γ to easily see that this expression approaches 1 as $\epsilon \to 0$. Thus, $\lim\limits_{\epsilon \to 0} R_m^*/R_m^N = 1$.

4. Griliches (1973) reports that estimates of the technological efficiency parameter tend to be about 0.1 for private research investments and 0.05 for public research investments in agriculture. Mansfield's (1965) highest estimate is 0.12.

5. More precisely, the credit is for R&D expenditures exceeding a base period level. Not all categories of expenditures (such as contract research) are completely credited, according to Mansfield.

6. The comparisons are not exact. In the uniform subsidy case, $\alpha_1 = \alpha_2 = \alpha = .05$. In the variable subsidy cases, $\alpha_1 \neq \alpha_2$. Either the difference is absolutely small (.0125 vs. .01), or the difference has an insignificant consequence on the results.

REFERENCES

Barzel, Yoram. 1968. "The Optimal Timing of Innovations," *Review of Economics and Statistics* (August).

Brown, Gardner, Jr., and Jon H. Goldstein. 1983. "A Model for Valuing Endangered Species," *Journal of Environmental Economics and Management* vol. 11, pp. 303–309.

Brown, Gardner M., Jr., and Joseph Swierzbinski. 1983. "Endangered Species, Genetic Capital and Cost-Reducing R and D," in D. O. Hall, N. Myers, and N. S. Margaris, eds., *Economics of Ecosystem Management* (Boston, Dr. W. Junk Publishers).

Ciriacy-Wantrup, S. V. 1952. *Resources Conservation* (Berkeley, University of California).

Dasgupta, Partha, and Joseph Stiglitz. 1980. "Industrial Structure and the Nature of Innovative Activity," *Economic Journal* vol. 80, pp. 266–293.

Griliches, Zvi. 1973. "Research Expenditures and Growth Accounting," in B. Williams, ed., *Science and Technology in Economic Growth* (New York, Wiley).

Leopold, Aldo. 1949. *A Sand County Almanac* (New York, Ballantine Books).

Mansfield, Edwin. 1965. "Technological Changes: Stimuli, Constraints, Returns," *American Economic Review* vol. 55 (May) pp. 310–322.

———. 1985. "Public Policy Toward Industrial Innovation: An International Study of Direct Tax Incentives for Research and Development," in K. Clark and colleagues, eds., *The Uneasy Alliance* (Boston, Harvard Business School Press).

Marglin, Stephen A. 1963. *Approaches to Dynamic Investment Planning* (Amsterdam, North-Holland).

Myers, Norman. 1983. *A Wealth of Wild Species* (Boulder, Colo., Westview Press).

Oldfield, Margery L. 1984. *The Value of Conserving Genetic Resources* (Washington, D.C., U.S. Department of the Interior, National Park Service).

Spence, A. Michael. 1984. "Cost Reduction, Competition and Industry Performance," *Econometrica* vol. 52 (January) pp. 101–121.

QUASI-OPTIMAL PRICING FOR COST RECOVERY IN MULTIPLE PURPOSE WATER RESOURCE PROJECTS

A. Myrick Freeman III
Jeffrey C. Norris

Much of the rigor and sophistication of present-day economic analysis of water resource development projects designed to serve multiple purposes can be attributed to the early work of John Krutilla and others in the 1950s and 1960s. John's own major work in this area includes his book with Otto Eckstein, *Multiple Purpose River Development: Studies in Applied Economic Analysis* (Krutilla and Eckstein, 1958); the report to the Bureau of the Budget on *Standards and Criteria for Formulating and Evaluating Federal Water Resources Development* (Hufschmidt and coauthors, 1961); and the influential discussion of issues in benefit and cost measurement—*Federal Natural Resources Development* (Krutilla and coauthors, 1969).

These works in resource and environmental economics played a major role in the early training of this paper's senior author. The occasion to return to this body of work for guidance in a theoretical area of current policy interest presented itself recently when he was asked to help prepare a research proposal to a public water resources agency seeking advice on how to design a system of prices or charges for its project outputs.

The agency requesting the assistance generally does not receive legislative appropriations, so in effect it operates under a budget constraint requiring self-financing of project developments. The theoretical question was how to design a system of prices when there were two possibly conflicting objectives to be served: (1) economic efficiency through rationing of project outputs where appropriate and (2) the raising of sufficient revenues to cover all or some specified portion of the total project costs. Efficiency generally calls for marginal cost pricing, except in the case of public goods. But where there is joint production, which is the essence of multiple purpose water resource development, marginal cost pricing might not generate the desired level of revenues. And a revenue shortfall is virtually certain where one or more of the project outputs is a public good.

Somewhat surprisingly, there seems to be no comprehensive treatment of the problem of conflicting pricing objectives in the early water resources literature, although there is some discussion of specific pricing problems. For example, Marglin (1962, p. 39), Krutilla (1969), Milliman (1969), and James and Lee (1971) discuss the problem posed by increasing returns to scale leading to a marginal cost price being below average cost. Krutilla (1969) and Marglin (1962, pp. 68–75) discuss the use of pricing policy to effect redistribution objectives. And there is much discussion of the problem of allocating joint costs where prices are to be set so as to recover fully allocated joint costs (Eckstein, 1958, pp. 259–272; Hirshleifer, DeHaven, and Milliman, 1960, pp. 93–94; Hufschmidt and coauthors, 1961, p. 56).

Krutilla (1966, 1969) and Milliman (1969), among others, discuss the inefficiencies that could arise from failure to set prices equal to marginal costs. And Krutilla (1969) points out that since prices determine the quantities demanded of project outputs, it is not possible to separate pricing decisions and cost recovery policy from project evaluation and design decisions. But we could find only two explicit references to the dilemma of conflicting pricing objectives when projects produce both marketable (or private) and nonmarketable (or public) goods as outputs. Although both references pose the problem in a manner consistent with the approach taken in this paper, neither contains a formal analysis of the problem.

In one of these two sources, Krutilla and Eckstein (1958) point to the likely example of the joint production of marketable hydropower and the public good of flood protection. On the pricing dilemma they write: "Unless the flood control costs could be grafted on to the inframarginal ranges of a set of discriminatory rate schedules, the added price on power to cover these costs would be similar to an excise [tax] on power and would adversely affect the marginal conditions for efficient distribution" (Krutilla and Eckstein, 1958, p. 192). Here the conflict between efficiency and cost recovery is explicitly recognized. And, of particular interest for our purposes, the equivalence of pricing above marginal cost and imposing an excise tax on the good is also recognized.

The other reference in the early literature to the problem of conflicting pricing objectives occurs in the discussion of the joint cost allocation problem in Eckstein (1958). Eckstein formulates the problem implicitly as one of second-best pricing. Citing articles by Manne (1952), Fleming (1953), and Boiteux (1956) on optimal pricing for multiproduct firms, he writes:

On grounds of pure economic theory, another principle has been suggested. It calls for minimizing the distortions in the price structure which inevitably result from the requirement that total revenues equal the total cost of reimbursable purposes rather than that price equals marginal cost. . . . In the general case, joint costs should be allocated in inverse proportion to the elasticities of demand of the outputs of the different purposes. (Eckstein, 1958, p. 268)

This is one version of the conditions for optimal pricing and joint cost allocation to be discussed in this paper. Insofar as we can determine, there have been no formal analyses based on this principle in the water resources pricing literature in the thirty years since Eckstein's book was published.

This paper first provides a formal characterization of the pricing problem facing a water resources development agency. It then shows that a now well-known line of analysis from the literatures of public utility pricing and public finance provides an integrated approach for the design of a pricing policy for multipurpose projects when the policy has the dual objective of raising revenue and achieving economic efficiency. This line of analysis goes back to Ramsey (1927) for optimal taxation and to Manne (1952), Fleming (1953), and Boiteux (1956) for optimal pricing of multiproduct firms.[1] More recently Baumol and Bradford (1970) brought together these several threads and discussed the four different versions of the conditions for quasi-optimal pricing or taxation— which we refer to hereafter as the Ramsey–Baumol–Bradford pricing rules (or RBB rules).

The first section below shows that the RBB rules provide a reconciliation of the efficiency and revenue-raising objectives of pricing policy in a second-best setting. The second section shows that the RBB rules have important implications for project design and the optimality conditions for the level of provision of public good outputs from water resource projects. As explained in the third section, the RBB rules imply that there is an efficient level of cost sharing between the purchasers of marketable outputs from the project and the public treasury. Finally, in the fourth section we show, as Eckstein recognized, that the RBB rules indirectly provide a formula for efficient allocation of joint costs—the efficient cost allocation being the one that would result in RBB prices if prices were determined according to fully allocated average costs.

No new theory is developed in this paper; its contribution instead lies in the integration of several strands from the existing public finance and public pricing literatures and in their application to the problem of pricing multiple purpose water resource project outputs, showing how several problems of concern in the earlier literature can be dealt with in a unified framework.

The Theory of Quasi-Optimal Pricing

Consider a river basin development agency that produces two marketable outputs X and Y and one public good output Z from water resource development projects in a river basin. The cost of these outputs is given by the cost function $C(X, Y, Z)$, which is increasing in each of its arguments. There are market demand functions for the two marketable outputs represented by

$$X^d = X^d(P^x, P^y, Z, M) \tag{1}$$
$$Y^d = Y^d(P^y, P^x, Z, M)$$

where M is the aggregate income of project customers. These are Marshallian demand functions since they determine how quantities demanded and project revenues vary with uncompensated changes in prices. Note that the output of the public good is an argument in the private good demand functions. There also is a total benefit function for the public good output $B(Z, X, Y, M)$.

The revenues of the agency depend upon the outputs of the two private goods and the prices charged to recipients of these goods. We assume throughout the following discussion that the agency sets prices for the private goods so as to clear the market, given output levels. Thus we can write the revenue function as $R(X, Y, Z)$. The level of the provision of the public good can affect revenue through its effects on the private demands and market-clearing prices. But provision of the public good produces no revenue directly. Also on the assumption that the agency cannot affect the aggregate income of consumers, M is suppressed as an argument in the revenue function. Finally profit, π, is given by $\pi = R - C$, which can be greater than, equal to, or less than zero.

The efficient, or Pareto-optimal, level of project outputs must satisfy the following conditions:

$$P^x = C_x$$
$$P^y = C_y \tag{2}$$
$$B_z = C_z$$

where subscripts indicate partial derivatives. We will denote the Pareto-optimal outputs as X^*, Y^*, and Z^*.

Nothing in this formulation of the problem would prevent the agency from realizing negative profits. Of course, there are costs associated with Z^* being greater than zero with no offsetting revenues due to the public good characteristic of Z. But even if $Z^* = 0$, profits could be negative because of increasing returns to scale or jointness in the cost function or both.

Suppose that the agency operates under a budget constraint:

$$\pi = R - C \geq S.$$

It is perhaps most convenient to think of the balanced budget case where $S = 0$. But if the general revenues are tapped to subsidize the provision of the public good output, S could be less than zero. Thus, the choice of S is a choice about the degree of cost sharing between local project beneficiaries and taxpayers in general. In this section, we take S as given exogenously. The choice of S is considered below.

The problem posed in this paper arises if

$$\pi(X^*, Y^*, Z^*) = R(X^*, Y^*, Z^*) - C(X^*, Y^*, Z^*) < S, \qquad (3)$$

for then the agency must raise the prices charged for the marketable outputs. Through the demand functions, the quantities demanded of these outputs are reduced and the conditions for an efficient set of output levels are violated. This is the problem that was analyzed by Baumol and Bradford (1970). It is a problem in the analysis of second best—or in the determination of quasi-optimal prices, to use their term. The rest of this section briefly reviews their analysis and the four alternative forms of the quasi-optimal pricing rule, assuming for the time being that $Z^* = 0$ and that X and Y are the only two goods produced in the economy. This latter assumption highlights the equivalence of the quasi-optimal pricing problem and the optimal commodity taxation problem.

Formally stated, the problem is to find that set of price increases above marginal costs (which are equivalent to commodity taxes) that minimizes the social cost of raising sufficient revenue to meet the profit constraint. Following Baumol and Bradford (1970), we take $W(P^x, P^y)$ as a consumer surplus measure of aggregate welfare. Maximizing W subject to the revenue constraint yields the following first-order conditions:

$$\frac{\partial W}{\partial P^x} = \lambda \frac{\partial \pi}{\partial P^x} \quad \text{and} \quad \frac{\partial W}{\partial P^y} = \lambda \frac{\partial \pi}{\partial P^y}, \qquad (4)$$

where λ is the shadow cost of raising the marginal dollar so as to just meet the revenue constraint. Since $\partial W/\partial P^x = -X$ (and similarly for Y), by substitution into equation (4) we obtain

$$\frac{-X}{\partial \pi/\partial P^x} = \lambda = \frac{-Y}{\partial \pi/\partial P^y}.$$

Rearranging this yields the first version of the RBB rule for quasi-optimal pricing:

I. The ratio between the marginal profit yields of unit changes in the *prices* of any two goods will be equal to the ratio between their output levels:

$$\frac{\partial \pi/\partial P^x}{\partial \pi/\partial P^y} = \frac{X}{Y}.$$

This rule holds in the general case when cross-elasticities of demand are nonzero.

Figure 1 provides a graphic analysis of the intuition behind rule I. For this example, assume that cross-elasticities of demand are zero. In the figure, the demand curves for X and Y are assumed to be the same, but the marginal cost of Y is assumed to be higher. Suppose that in an effort to raise additional revenue, the agency raises the price of X above C_x to $P^{x'}$. The marginal profit yield for a change in price is given by

$$\frac{\partial \pi}{\partial P^x} = P^x \frac{\partial X}{\partial P^x} + X - C_x \frac{\partial X}{\partial P^x}$$

$$= X + (P^x - C_x)\frac{\partial X}{\partial P^x}.$$

(5)

At these prices, the marginal profit yields for X and Y are the distances AB minus CD and AB, respectively. Thus, rule I is violated:

$$\frac{\partial \pi / \partial P^x}{\partial \pi / \partial P^y} < \frac{X}{Y}.$$

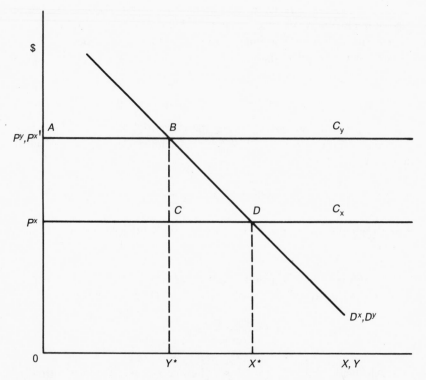

Figure 1. Analysis of intuition underlying quasi-optimal pricing rule I.

At this point, since the social welfare costs of one-unit price changes are the same for both X and Y, revenue can be increased by the distance CD through a one-unit increase in P^y and a one-unit decrease in P^x holding social welfare constant. Alternatively, a one-unit increase in P^y will allow a more-than-one-unit decrease in P^x with no net change in total revenue. Yet social welfare is increased because of the larger welfare gain associated with the larger decrease in P^x. Such adjustments should continue until rule I is satisfied.

The second version of the RBB rule is derived by substituting the expression for marginal profit yield, equation (5), into equation (4) and rearranging to obtain

$$-X\frac{\partial P^x}{\partial X} = \lambda(P^x + X\frac{\partial P^x}{\partial X} - C_x).$$

Then, by adding $P^x + X\dfrac{\partial P^x}{\partial X} - C^x$ to both sides and recalling that $MR^x = P^x + X\dfrac{\partial P^x}{\partial X}$, we get:

II. For each output, the deviation of quasi-optimal price from marginal cost must be proportionate to the difference between the output's marginal cost and marginal revenue.[2]

$$P^x - C_x = (1 + \lambda)(MR^x - C_x)$$

or

$$\frac{P^x - C_x}{MR^x - C_x} = 1 + \lambda$$

This rule holds only when all cross-elasticities of demand are equal to zero.[3]

At this point we can establish bounds on the sign and magnitude of λ—the shadow cost of the profit constraint. From equation (4), it is clear that λ must be negative. And since at the quasi-optimal price, marginal revenue must be less than marginal cost (otherwise profits could not be increased by reducing output), then $\lambda < -1$.

For the third version of the rule, rearrange rule II to obtain

$$-\lambda(P^x - C_x) = (1 + \lambda)X\frac{\partial P^x}{\partial X}. \tag{6}$$

Defining the price elasticity of demand as

$$E^x = -\frac{\partial X}{\partial P^x} \cdot \frac{P^x}{X},$$

divide equation (6) by $X \cdot P^x$ to obtain rule III:

III. For each output, the percentage deviation of quasi-optimal price from marginal cost must be inversely proportionate to its price elasticity of demand:

$$\frac{P^x - C_x}{P^x} = \frac{1 + \lambda}{\lambda} \cdot \frac{1}{E^x}.$$

This condition holds only when all cross-elasticities of demand are equal to zero.

To obtain the fourth version of the rule, let ΔP^x represent the quasi-optimal change in price or the tax. Using this, rule III can be rearranged to yield

$$\frac{\partial X}{\partial P^x} \Delta P^x = \frac{1 + \lambda}{\lambda} \cdot X.$$

And since $\frac{\partial X}{\partial P^x} \Delta P^x$ approximates ΔX, we obtain rule IV:

IV. Quasi-optimal prices must yield outputs that deviate by (approximately) the same proportion from those which would result from pricing at the marginal costs corresponding to the quasi-optimal output levels:

$$\Delta X = kX$$

where $k = \dfrac{1 + \lambda}{\lambda}$.

This rule also holds in the general case when cross-elasticities of demand are nonzero.

The four rules discussed here determine the *structure* of RBB prices. The factors influencing the optimal *level* of RBB prices are considered next. The level of RBB prices is determined by the choice of S, the budget constraint on the agency.

The Level of Quasi-Optimal Prices

It will be recalled that in the derivation of the four RBB pricing rules, it was assumed that the river basin agency and the economy were coextensive: that is, the agency had the power to set prices on all of the goods in the economy so that it could spread the social cost of price distortions in a quasi-optimal manner over all marketable goods being produced and sold. Now suppose that the agency produces and sells the two marketable outputs, X and Y, and that there is a competitive sector in the economy producing a vector of private goods. Further suppose that while the agency operates under a binding budget constraint, there is a political constraint preventing the imposition of taxes on the competitive sector. This is a problem of *third*-best, because of the additional constraint. The agency should still follow the RBB pricing rules for those outputs that it produces; given the constraint against taxing the competitive sector, there is nothing better it can do. But the social costs of the distortions resulting from taxing only project outputs will be much higher than they would be if the agency could meet part of its revenue requirements through taxes on the other goods. Thus, there is an efficiency argument for cost sharing even when all of the agency's outputs are marketable. (There may, of course, also be equity or redistribution objectives to be served by cost sharing, but those issues are beyond the scope of this paper.)

The RBB pricing rules provide a basis for determining the efficient level of cost sharing, that is, the efficient division of total project costs between the purchasers of the project's marketable outputs and taxpayers. The agency should optimally distort its prices upward until it reaches the point where the social cost of raising an additional dollar of revenue through RBB pricing just equals the marginal social cost of raising one dollar of revenue from taxpayers. Beyond that point, if the marginal social cost of tax dollars is constant, all additional revenues should come from general taxation. Otherwise, both project prices and federal taxes should be adjusted upward while maintaining the equality of the marginal social costs of the two sources of revenue. Rough estimates of the marginal social cost of federal tax revenues are available in the literature (see, for example, Browning [1976, 1987]; Ballard, Shoven, and Whalley [1985]; and Wildasin [1984]).

There are three ways of determining the marginal social costs of revenues from quasi-optimal project pricing. The first follows from the original formulation of the problem as one of maximizing a consumer surplus measure subject to the agency's budget contraint. The Lagrangian multiplier gives the social cost of raising the budget constraint by one dollar in terms of lost consumer surplus. Thus, marginal social cost of an additional dollar is equal to $-\lambda$.

Another way to see this is to recall that the lost consumers' surplus for a one-unit increase in the price of, say X, is given by X. This increase in P^x produces $\partial \pi / \partial P^x$ in revenues. So the lost consumer surplus per dollar can be measured by

$$\frac{X}{\partial \pi / \partial P^x},$$

which is equal to $-\lambda$, as shown above. This measure can be used when cross-elasticities of demand are nonzero.

Finally, if the output of X is reduced by one unit, the lost consumer surplus is equal to $P^x - C_x$ while the revenue raised is equal to $-(MR^x - C_x)$. Thus, the lost consumer surplus per dollar of revenue is

$$- \frac{P^x - C_x}{MR^x - C_x},$$

and the social cost of the dollar raised is greater by this amount. So the social cost of a dollar of revenue is

$$1 - \frac{(P^x - C_x)}{(MR^x - C_x)},$$

which from rule II is equal to $-\lambda$. This measure is valid only in the case of independent demands.

Now let us return to the general case, in which the agency produces a public good Z as well as marketable outputs. The agency incurs additional costs in the production of the public good but receives no additional revenue because of its inability to charge a price for Z. Again, the most efficient way to raise the necessary revenue is to optimally distort the prices of marketable outputs according to the RBB rules and, if possible, to increase the level of cost sharing from the general tax revenues as necessary to keep the marginal social costs of revenues from all sources equal. The lower the marginal social cost of tax revenues, the greater is the optimal degree of cost sharing and the smaller the deviation of RBB pricing from first-best marginal cost pricing.

Implications for Project Design

As the preceding analysis makes clear, RBB pricing for cost recovery has implications for the choice of output levels for marketable goods. What has not been generally recognized in the water resources literature is that RBB pricing also has implications for the level of public good provision from water resource projects. The conventional wisdom still appears to be that the condition requiring that marginal benefit equal marginal cost should apply for public goods from water resource projects (see, for example, Young and Haveman [1985], p. 478). But a marginal change in the level of public good provision entails not only an increase in the real resource cost of production as measured by C_z but also an increase in the distortionary cost of raising the revenues to pay for the additional resources. As shown in the preceding section, the true social cost of raising the revenue from RBB pricing is given by $-\lambda$. Thus, the condition for the efficient or quasi-optimal level of provision of the public good is $B_z = -\lambda C_z$.[4]

This simple statement of the quasi-optimal condition for public good provision is strictly true if and only if the level of the public good provision does not affect the private goods demand curves. In our original statement of the problem, we assumed that Z was an argument in the private goods demand functions. A marginal increase in Z could shift the demand curves for X and Y in such a way as to change their prices and quantities. Thus, the profits from distortionary taxes could increase or decrease depending on the relative magnitude of the shifts in the demand curves. If the profits increased (decreased), the marginal social cost of RBB pricing to recover the incremental cost of Z would be lower (higher) than indicated above. Wildasin (1984) has derived a general equilibrium measure of the efficiency condition for public goods provision which incorporates both the social cost of distortionary taxation and the effects of public good provision on private goods demand and tax revenue. He does not assume quasi-optimal pricing or taxation and considers a tax on only one private good. But the extension to many goods is straightforward. In our notation and assuming that the extra revenue is raised through quasi-optimal pricing, the efficient level of public good provision is where

$$B_z = -\lambda \left[C_z - (P^x - C_x)\frac{\partial X}{\partial Z} - (P^y - C_y)\frac{\partial Y}{\partial Z} \right].$$

Conceivably the right-hand side of this expression could be less than C_z. This would require a relatively strong net revenue-enhancing effect of the public good on private good demands and a value for λ close to 1 in absolute value. Given these conditions, the optimal level of the public good would be greater than that indicated by the first-best, marginal-benefit-equals-marginal-cost condition. Finally, it should be noted that if there is cost sharing, the preceding expression must be modified to include a term reflecting the effect of RBB pricing or taxation on the revenues collected from nonproject market goods as well.

Implications for Joint Cost Allocation

There is a substantial literature on the allocation of the joint costs of water resources development to specific purposes.[5] The purpose of joint cost allocation is to find a price for each of the marketable outputs of the project that will lead to the recovery of the full cost of producing each output—where full cost includes a "proper" proportion of the joint costs of the various outputs. The procedures involve setting prices equal to some measure of average costs that includes both separable and joint costs of production. There is general agreement in this literature that there is no logically correct method for allocating joint costs and that any allocation is essentially arbitrary.

The theory of RBB pricing provides another way of looking at the problems of joint cost allocation and cost recovery.[6] The RBB rules yield a set of prices for marketable outputs that will recover the politically determined portion of total project costs, including, presumably, joint costs. There is an allocation of the joint costs that, if used to determine a set of prices in the conventional manner, will reproduce the RBB prices. This "allocation" of joint costs will be efficient in the sense of minimizing the distortions of nonmarginal cost pricing.

In the simple case of independent demands, this joint cost allocation can be determined in the following manner. First, the joint cost allocated to X is used to set P^x above C_x. Thus, the cost allocated to output X must satisfy the following condition:

$$\frac{AJC^x}{X} = P^x - C_x,$$

where AJC^x is the joint cost to be allocated to the production of X and P^x is the quasi-optimal price of X. This expression can be substituted in RBB rule III to solve for AJC^x:

$$AJC^x = \left(\frac{1 + \lambda}{\lambda}\right)\frac{P^x \cdot X}{E^x}.$$

It should be noted that the sum of the costs allocated to marketable outputs by this expression bears no logical relationship to joint costs as conventionally defined. Rather, the sum is determined by the budget constraint imposed on the agency—which means that this expression is not really a cost allocation in the conventional sense of the word but is instead a procedure for making the RBB prices appear to cover the average costs of marketable outputs.

Conclusions

As shown in this paper, the Ramsey–Baumol–Bradford (RBB) pricing theory provides a consistent framework for dealing with several issues in the economics of multiple purpose water resources development. It provides an efficient set of prices for meeting a budget constraint imposed upon a public water resources agency. It also provides a way of looking at the political issues of cost sharing and joint cost allocation. And it shows that the requirements of project finance in terms of cost recovery and revenue raising have implications for project design and choice of output levels for both the marketable and the public goods outputs of the project. Although our exposition has been framed in terms of multiple purpose water resources development, the same principles apply to the management of other resources yielding multiple outputs—for example, public lands.

A government agency wishing to employ these principles in determining a set of prices for a project's marketable outputs would have to proceed along the following lines. A key decision concerns the level of cost recovery through pricing, or the choice of S. This determines the level of RBB prices and therefore the marginal social cost of the distortionary prices. This choice could be made on efficiency grounds, as described in the section above on "The Level of Quasi-Optimal Prices," or it could reflect concern for the distribution of welfare between consumers of the marketable outputs and taxpayers. Then, in addition to using the RBB rules to determine the structure of prices, the agency must use the data on the demand for the private goods and price distortions to determine the quasi-optimal level of the public good. And finally, as John Krutilla (1966, 1969) pointed out many years ago, there must be consistency between the prices that are actually charged to finance the project and to ration the outputs and the prices assumed for the purpose of project evaluation or benefit–cost analysis.

Acknowledgments

Some of the results in this paper were first reported by Jeffrey Norris in his senior honors thesis, Bowdoin College, May 1986. The authors are indebted to John Fitzgerald and V. Kerry Smith for helpful comments.

NOTES

1. See Bös (1985) for a recent review of this literature.

2. Of course, similar versions of this rule and of rules III and IV, presented below, can be derived for all of the other project outputs.

3. For a formal derivation of the pricing rules with interdependent demands, see Bös (1985, pp. 151–152).

4. There is a hint that Eckstein (1958, p. 268) recognized this point. The point is well known in the modern public finance literature; see, for example, Atkinson and Stiglitz (1980, pp. 490–492), or Auerbach (1985, pp. 110–112).

5. See, for example, Eckstein (1958, pp. 259–272); Hirshleifer, DeHaven, and Milliman (1960, pp. 93–94); and Loughlin (1977). For examples of more recent contributions employing game-theoretic concepts, see Heaney and Dickinson (1982); and Young, Okada, and Hashimoto (1982).

6. This was recognized some years ago by Michael D. Bowes and John V. Krutilla in a comment published in the *Journal of Forestry* (Bowes and Krutilla, 1979), as we found out after completing the first draft of this paper. Baumol,

however, denies that RBB pricing is a form of cost allocation, saying that to make that claim involves "a play on words" (Baumol, 1986, pp. 143–144). To him, a cost allocation is an "a priori formula based *only* on costs and output quantities. . . ." But Baumol's point seems to us to be based on an artificial distinction, since as a practical matter cost allocations are important because of their use in setting prices.

REFERENCES

Atkinson, Anthony B., and Joseph E. Stiglitz. 1980. *Lectures on Public Economics* (New York, McGraw-Hill).

Auerbach, Alan J. 1985. "Excess Burden and Optimal Taxation," in Alan J. Auerbach and Martin Feldstein, eds., *Handbook of Public Economics* vol. I (Amsterdam, North-Holland).

Ballard, Charles L., John B. Shoven, and John Whalley. 1985. "General Equilibrium Computations of the Marginal Welfare Costs of Taxes in the United States," *American Economic Review* vol. 75, pp. 128–138.

Baumol, William J. 1986. *Superfairness: Applications and Theory* (Cambridge, Mass., MIT Press).

———, and David F. Bradford. 1970. "Optimal Departures from Marginal Cost Pricing," *American Economic Review* vol. 60, pp. 265–283.

Boiteux, M. 1956. "Sur la gestion des Monopoles Publics astreints a l'equilibre budgetaire," *Econometrica* vol. 24, pp. 22–40.

Bös, Dieter. 1985. "Public Sector Pricing," in Alan J. Auerbach and Martin Feldstein, eds., *Handbook of Public Economics* (Amsterdam, North-Holland).

Bowes, Michael D., and John V. Krutilla. 1979. "Cost Allocation in Efficient Multiple-Use Management: A Comment," *Journal of Forestry* vol. 77, pp. 419–420.

Browning, Edgar K. 1976. "The Marginal Cost of Public Funds," *Journal of Political Economy* vol. 84, pp. 283–298.

———.1987. "On the Marginal Welfare Cost of Taxation," *American Economic Review* vol. 77 (March) pp. 11–23.

Eckstein, Otto. 1958. *Water-Resource Development: The Economics of Project Evaluation* (Cambridge, Mass., Harvard University Press).

Fleming, J. M. 1953. "Optimal Production with Fixed Profits," *Economica* N.S. vol. 20, pp. 215–236.

Heaney, James P., and Robert E. Dickinson. 1982. "Methods for Apportioning the Cost of a Water Resource Project," *Water Resources Research* vol. 18, pp. 476–482.

Hirshleifer, Jack, James C. DeHaven, and Jerome W. Milliman. 1960. *Water Supply: Economics, Technology, and Policy* (Chicago, University of Chicago Press).

Hufschmidt, Maynard, John Krutilla, and Julius Margolis, with Stephen Marglin. 1961. *Standards and Criteria for Formulating and Evaluating Federal Water Resources Development*. Report of the Panel of Consultants to the Bureau of the Budget (Washington, D.C., June 30).

James, L. Douglas, and Richard R. Lee. 1971. *Economics of Water Resources Planning* (New York, McGraw-Hill).

Krutilla, John V. 1966. "Is Public Intervention in Water Resources Development Conducive to Economic Efficiency?" *Natural Resources Journal* vol. 6, pp. 60–75.

————. 1969. "Efficiency Goals, Market Failure, and the Substitution of Public for Private Action," in U.S. Congress, Joint Economic Committee, *The Analysis and Evaluation of Public Expenditures: The PPB System* vol. I, pp. 277–289.

————, and Otto Eckstein. 1958. *Multiple Purpose River Development: Studies in Applied Economic Analysis* (Baltimore, Md., Johns Hopkins Press for Resources for the Future).

————, Jack L. Knetsch, Robert H. Haveman, Charles W. Howe, and Michael Brewer. 1969. *Federal Natural Resources Development: Basic Issues in Benefit–Cost Measurement* (Washington, D.C., George Washington University, Natural Resources Policy Center).

Loughlin, James C. 1977. "The Efficiency and Equity of Cost Allocation Methods for Multi Purpose Water Projects," *Water Resources Research* vol. 13, pp. 8–14.

Manne, A. S. 1952. "Multiple-Purpose Enterprises—Criteria for Pricing," *Economica* N.S. vol. 19, pp. 322–326.

Marglin, Stephen A. 1962. "Objectives of Water-Resource Development: A General Statement," in Arthur Maass and others, eds., *Design of Water-Resource Systems: New Techniques for Relating Economic Objectives, Engineering Analysis, and Governmental Planning* (Cambridge, Mass., Harvard University Press).

Milliman, Jerome W. 1969. "Beneficiary Charges and Efficient Public Expenditure Decisions," in U.S. Congress, Joint Economic Committee, *The Analysis and Evaluation of Public Expenditures: The PPB System* vol. I, pp. 291–318.

Ramsey, Frank P. 1927. "A Contribution to the Theory of Taxation," *Economic Journal* vol. 37, pp. 47–61.

Wildasin, David E. 1984. "On Public Good Provision with Distortionary Taxation," *Economic Inquiry* vol. 22, pp. 227–243.

Young, Robert A., and Robert H. Haveman. 1985. "Economics of Water Resources: A Survey," in Allen V. Kneese and James L. Sweeney, eds., *Handbook of Natural Resources and Energy Economics* vol. II (Amsterdam, North-Holland).

Young, H. P., N. Okada, and T. Hashimoto. 1982. "Cost Allocation," *Water Resources Research* vol. 18, pp. 463–475.

BALANCING MARKET AND NONMARKET OUTPUTS ON PUBLIC FOREST LANDS

William F. Hyde
Steven E. Daniels

Public ownership of forest lands in the United States must be viewed as a market intervention of major importance considering that public agencies manage 482.5 million acres, or 28 percent, of the commercial forest land in this country (U.S. Department of Agriculture, Forest Service, 1982, p. 349). While there are well-established justifications for public ownership both in concept and in practice (see, for example, Krutilla and Eckstein [1958], pp. 41–58), this paper is concerned less with the justification than with the analysis of the impact of public ownership. The analysis itself is important because the literature thus far has not dealt with the critical issue of production costs. Our evaluation of the effects of public ownership, which suggests that there are differences in outcome depending upon the relative weight given to market-responsive or efficiency-based policies, invites inquiry into the implications of these differences for public management. Therefore, the primary question addressed here is whether regulations that restrict the responses of public land managers to market signals affect social welfare positively or negatively. A further question is whether these regulations improve distributive justice.

Since it is not possible to examine all of the restrictions on public land managers, these questions are considered within the framework of two illustrative cases involving the management of goods and services from the public lands. The first case deals with the marketing of public timber resources and the second with recreation management.

In the first case, although timber is a good that the market values accurately, its output is severely constrained by various regulations adopted by public forest land agencies. These regulations impose constraints on managerial responses both to market timber prices and to the costs of timber production and timber sale preparation. Previous studies have examined the impacts of the price constraints but not of cost constraints. Considerable efficiency gains would appear to result from reforms relaxing both the price and the cost constraints.

In addition to commodities such as timber, the public lands also provide many non-market-valued goods and services. Indeed, both the historical genesis and much of today's justification for the existence of the public lands have been that nonmarket goods and services are too important (in terms of consumers' values) to ignore. Perhaps the most obvious case in point today is recreation.

The difficulty of excluding recreational users from most forest lands implies that many forms of recreation have attributes of public goods. It does not, however, remove all market and market-proxy data from the knowledge of forest managers. The second case presented in this paper considers the efficiency gains and allocative differences that might result based on the data and managerial skills that are currently available if the U.S. Forest Service campgrounds in a particular valley in western Montana were managed in accordance with efficiency criteria.

As with the studies to date on public timber management, the economics literature thus far has considered the price responsiveness but not the cost component of recreation. Efficiency recommends the use of an allocative solution that sets marginal social benefits from recreation equal to marginal costs. Estimating the marginal social benefits raises questions in two important areas—(1) nonmarket valuation and (2) the effects of budget limitations on the provision of forest recreation—that this paper addresses. Because the efficiency approach presented here is generalizable to other forms of recreation, to other public agencies, and to other locations, the results presented below for the western Montana valley have more general implications.

Three hypotheses emerge from these case studies: (1) there is much gained from straightforward analysis with existing data; (2) the skills for inquiring into these timely policy questions already exist within the agencies; and (3) budget restrictions keep agencies lean and approximately efficient. The sections that follow examine the timber and recreation examples in turn, and the paper concludes with a summary statement on efficiency criteria and public land management.

Public Timber Management

Harvest Decision Criteria

The current U.S. Forest Service criteria for decisions on the harvesting of the public timber resource are similar to the criteria of the federal Bureau of Land Management and Bureau of Indian Affairs and of many state agencies, including the largest of the state forest management agencies—the Washington State Department of Natural Resources. (This similarity is one reason for building the analysis presented here on a foundation of U.S. Forest Service timber management planning.)

After a brief description of the harvest decision criteria, the first subsection below considers one method for measuring the substantial variation from market efficiency results for timber management that these criteria cause. The second subsection discusses the results of our measurement in both efficiency and distributive terms.

The public timber management agencies, both federal and state, produce 22 percent of the annual softwood harvests in the United States, with the national forests alone producing 15 percent of these harvests (U.S. Department of Agriculture, Forest Service, 1982, p. 424). (The national forest share of total public softwood production provides additional justification for basing this analysis on U.S. Forest Service timber management planning.) On a regional basis the public-sector share of total softwood production ranges from 66 percent in the Rocky Mountains to 6 percent in the Northeast. Therefore, it would be expected that the regional impact of public management could range from substantial to inconsequential.

The decision regarding the annual timber volume offered for sale by the U.S. Forest Service originates from the ten-year plan established for each of the country's 154 national forests. A constrained-volume maximization rule determines the size of the planned harvests. Its objective is the maximization of sawtimber volume over time given the current timber stand ages, volumes, and anticipated growth. The Forest Service constrains the solution volume so that planned harvests in future decades never fall below those of the current decade, although planned annual harvests may fluctuate within the range of plus or minus 10 percent from one-tenth of the ten-year total. Neither production costs nor market prices play a role in this harvest calculation. Clearly, then, we anticipate that the timber volumes offered for sale by the Forest Service will be quite different from those determined by market efficiency.

Analytical Approach

The economics literature generally models public agency timber supply as fixed at its predetermined, planned level. This is not wholly consistent with reality, but it is consistent with the Forest Service objective of producing a constant volume of timber regardless of market conditions. The assumption of fixed supply departs from reality in that the offered volume is not always either bid upon or sold; furthermore, even if the offered volume is sold, the logger with the winning bid may delay the harvest operation for as many as five years while waiting for better market conditions. Nevertheless, our analysis is as much concerned with the advisability of pursuing the Forest Service objective as with its previous success. Therefore, we remain consistent with the majority of the literature and take public agency production as fixed and invariant with the market. Our inquiry investigates how public agency success at meeting the objective of fixed timber harvest volume varies from social efficiency.

Our procedure begins with the choice of 1977 as the year for comparative market analysis because it is the most recent year for which we can be confident of the coincidence of our data with the needs of our analytical mode. We examine supply and demand for softwood sawtimber under current harvest decision criteria and contrast our results with those occurring under a simulated alternative scenario in which the public sector follows market decision criteria. We are interested in the contrasting market clearing prices, in the public- and private-sector harvest levels and relative producer surpluses, and in a measure of contrasting consumer welfare. Regional knowledge of contrasting market equilibria permits regional as well as national estimates of these differential impacts on consumers and public and private producers.

It should be emphasized that this approach yields results that reflect one-time (1977) effects only. They suggest—but only suggest—directions of change for longer-term effects. A true longer-run analysis would require dynamic linkages of both the demand and supply models through time as well as investment and inventory feedbacks from supplies of one year to those of subsequent years.

Our empirical measures rely heavily on the aggregate timber modeling of Adams and Haynes (1981) as revised (D. M. Adams, personal communication, 1985). The Adams–Haynes (AH) model is the best-known econometric model of the forestry sector and the only one in the public domain. It is also the model that the Forest Service uses for its projections of the aggregate timber economy. The AH model provides softwood stumpage supply functions for eight regions and two private sectors—the forest industry and other private forest landowners. Our assessment of the impact of current public timber management policies follows the AH model in initially taking public supply as fixed at the observed 1977 level.

Our estimate of total public timber management and harvest costs equals the sum of the 1977 Forest Service timber management budget, the budget for reforestation as required by the Knutson–Vandenberg Act of 1930, and one-half of the road and trail budget plus 10 percent overhead. This sum is adjusted upward proportionally for the non–Forest Service share of 1977 public harvests. The final cost estimate is conservative to the extent that other Forest Service activities (e.g., fire control, insect and disease protection, some research) are actually inputs to timber management. It is also conservative to the extent that the Forest Service budget is largely an input to anticipated harvests many years in the future, although we compare this current budget with current receipts, not with expected future receipts. (If expenditures and receipts remain constant over time, this is equivalent to assuming a zero discount rate.[1])

Adams and Haynes (1981) derive stumpage demand for all but two regions from the demand for lumber and wood products, and they incorporate in their model a sophisticated understanding of the opportunities for interregional flows of various intermediate products. Because stumpage demand from the Northeast and from the North Central regions is small, Adams and Haynes treat it as a constant.[2] In our analysis we disregard these two regions because their public harvest share is insignificant at our level of detail.

Equilibrating the AH-derived demand functions from the remaining six regions with the AH private-sector supply functions and the exogenously fixed public harvests provides a basis for reasonable estimates of producer and consumer welfare arising from current public timber management practice. All of the functions are linear in form. Adams and Haynes choose between the basic specifications for their supply functions; their criterion is the best statistical fit. Tables 1 and 2 (presenting base case stumpage demand and base case stumpage supply, respectively) summarize the functions underlying these calculations. These tables continue the AH pattern of reference to stumpage prices in 1967 dollars per thousand board-feet (Mbf) and output in millions of cubic feet (MMcf). (The use of 1967 dollars by Adams and Haynes means that all of our final observations which occur in value terms must be inflated to 1977 using the same producer price index that these researchers used.)

Let us turn now to the simulations of market-sensitive public supply. Deriving the public supply functions for each region is a difficult problem, and our first-generation effort is admittedly ad hoc. We are looking for supply functions that, like the AH private-sector functions, reflect the costs of growing timber and preparing it for harvest. This suggests beginning with the AH direct supply functions (from table 2) and transforming them into indirect functions that display price intercepts and price (cost) responses to harvest and inventory levels. The new problem is to modify these functions until their intercepts and the coefficients on their response terms resemble those of cost-sensitive public functions. We consider adjustments for the intercepts and coefficients in turn.

Table 1. Base Case—Stumpage Demand

Derived demand equation: $Q = Q_d(P)$

Region	Intercept	Price (1967$/Mbf)
Southeast	912.8	−1.5466
South Central	1,554.1	−3.0857
Rocky Mountains	864.2	−0.1426
Pacific Southwest	798.0	−2.5127
Pacific Northwest (West)	2,795.6	−4.5153
Pacific Northwest (East)	561.5	−1.7226

Note: Q = quantity harvested in millions of cubic feet; P = stumpage price (1967 $/Mbf); Mbf = thousand board-feet.

Source: The results in this table are derived from Forest Survey data and an application of the Adams–Haynes (1981) model, as revised in 1985, with our modifications as discussed in the text of this paper.

Table 2. Base Case—Stumpage Supply

Supply equations. Private sector: $Q = Q_s(P, I)$ or $Q/I = Q_s(P)$

Public sector: $Q = \overline{Q}_s$

Region/ landowner	Dependent variable	Intercept	Price (1967$/Mbf)	Inventory (MMcf)	Trend (1/t)
Southeast					
FI	Q	−434.517	4.03927	0.049082	
OP	Q	273.313	2.96201	0.002691	
Public harvests = 57.79 MMcf; costs = $11,697,237					
South Central					
FI	Q	−434.517	4.03927	0.0949082	
OP	Q	273.313	2.96201	0.006329	
Public harvests = 127.93 MMcf; costs = $17,335,861					
Rocky Mountains					
FI and OP	Q/I	0.0093917	0.0003513		
Public harvests = 546.92 MMcf; costs = $54,192,925					
Pacific Southwest					
FI	Q/I	0.0252454	0.0004464		
OP	Q	−1,863.03	0.732612	0.108447	69,250.9
Public harvests = 369.37 MMcf; costs = $27,566.929					
Pacific Northwest (West)					
FI	Q/I	0.024569	0.0006436		
OP	Q/I	−0.0081291	0.0007948		
Public harvests = 1,032.17 MMcf; costs = $56,695,406					
Pacific Northwest (East)					
FI	Q/I	0.0252322	0.0000775		
OP	Q/I	−0.0007838	0.000287		
Public harvests = 370.01 MMcf; costs = $23,496,576					

Notes: FI = Forest industry; OP = other private forest landowners; Q = quantity harvested (MMcf); P = stumpage price (1967$/Mbf); and I = stumpage inventory (MMcf). Mbf = thousand board-feet; MMcf = million cubic feet.

Source: The results in this table are derived from Forest Survey data and an application of the Adams–Haynes (1981) model, as revised in 1985, with our modifications as discussed in the text of this paper.

The scant evidence available suggests a public equilibrium price at a level perhaps $20 per thousand board feet (in 1977 dollars) higher than private stumpage prices. This price difference may be due to higher public timber production costs, including the costs of managing a large bureaucracy, that are not replicated in smaller private operations; or, it may be due to multiple use constraints or even to something else we cannot identify. For whatever reason, a price differential apparently exists. The level of the public supply function, and therefore the price intercept, must be adjusted accordingly.[3]

The public-sector response to harvest quantity changes might reasonably mimic marginal production costs for the more similar of the two private sectors. For this reason, we identify the private sector in each region that most closely resembles that region's public sector in land quality and market share. We then choose the AH specification of its supply function for the underlying function from which we make (intercept and coefficient) adjustments to obtain a public supply function. This AH function for each region provides the coefficient of price response to harvest changes.

Finding an appropriate public inventory coefficient is a more complex problem. Inventory is exogenous in the AH supply functions, but the potentially harvestable inventory is actually a function of market price. This point is less important for the stable market conditions that Adams and Haynes model, but it becomes very important if that market suddenly destabilizes. Throwing a large public inventory on the market is just what demand-sensitive public harvest criteria recommend. Therefore, our simulation must make two inventory adjustments: one that reflects the expected relative differences between the public sector and the mimic private sector with respect to both inventory per acre and inventory quality, and another that adjusts the measure of public inventory itself to reflect the equilibrium price level.

The first adjustment in the inventory coefficient reflects the relative differences between public- and mimic private-sector inventory per acre and inventory quality. Harvest preparation costs per unit of volume decrease in response to improvements in each of these inventory measures. Per acre inventory data are readily available from the Forest Service (USDA, Forest Service, 1978). Site-class midpoints are our indicators of inventory quality, the basis of our argument being that site class is a measure of biological quality, and, in general, lands of poorer biological quality may be more poorly managed and less frequently harvested than lands of better quality. Therefore, a larger share of the standing timber on lands of poor quality may be decaying, inadequately spaced, and so forth, all of which creates poorer quality. The full adjustment multiplies the mimic forest industry (FI) or other private (OP) inventory coefficient by the following:

$$\frac{\text{public inv./acre}}{\text{FI-OP inv./acre}} \times \frac{\text{quality adj. pub. inv.}}{\text{quality adj. FI-OP inv.}}.$$

The final adjustment converts inventory from a purely physical measure into a term with economic meaning. Marginal stands (and marginal forest lands) now (implicitly) enter the inventory when stumpage prices increase and depart when stumpage prices decrease. Even if the Forest Survey inventory measure does not reflect this adjustment, the bidding of those loggers who compete for public timber sales does. We assume that the physical and economic inventories are in balance for the 1977 base year under current management. There is a direct relationship between economic inventory and price, and the greater the difference in simulated and actual equilibrium prices the greater the difference in economic inventories. Following this reasoning, our final adjustment multiplies the 1977 Forest Survey inventory for public lands by the ratio of the eventual simulated equilibrium price to the actual 1977 AH equilibrium price.

In order to compare our new simulated public supply functions with the AH functions, our functions must be transformed back to direct supply functions. Table 3 shows these simulated market-responsive public supply functions and the new regional AH-derived demand functions that appear in response to them.[4] Equilibrating the supply-and-demand functions and contrasting the results with those obtained for the base case permit one estimate of the welfare effects arising from market-responsive public softwood timber supplies. It must be noted that derivation of our simulated public supplies generated more confidence with regard to the direction of changes in welfare effects than with regard to the absolute levels of these effects; nevertheless, we show welfare effects as indicative of the magnitude of importance of this policy issue.

Results

Tables 4 and 5 show our estimates for the 1977 price and quantity changes and the consumers' and producers' surplus changes that result from introducing market-sensitive public timber harvest criteria. Our results are conjectural— and depend heavily on both the inventory term and the decrease in public production costs, but we expect no debate over the direction or magnitude of these terms. Therefore, we think that the price, quantity, and surplus measures in tables 4 and 5 are also of correct sign and magnitude.

Our greatest confidence is in the results for the four western regions: the public presence is greatest in these areas; in addition, Adams and Haynes have the greatest confidence in their own model for these regions. The public producers' surpluses for the Pacific Southwest and the Pacific Northwest (East), on the other hand, are ambiguous. (That is, public supply in these regions is highly price inelastic. These inelasticities prevent the model from finding a stable solution in a small number of iterations.) Both the directions and rank orders of all other changes conform with expectations based on the size of existing public inventories and their accessibility.

As table 4 shows, the most notable shifts are the southern increases in public-sector harvests (explained by existing large inventories and long rotations) and the western increases in public-sector harvests and relatively comparable decreases in combined forest industry and other private harvests. The only truly substantial western shift, however, is the very large public harvest increase (and corresponding forest industry decrease) in the Douglas fir region (Pacific Northwest [West]). There are stumpage price decreases in all of the western regions except the Rocky Mountains. The signs and orders of mag-

Table 3. Fully-Market-Responsive Stumpage Demand[a] and Supply[b]

Region	Intercept	Price (1967$/Mbf)	Public supply equations
Southeast	904.8	−1.5466	$Q = 273.313 + (P - 10.299)(2.96201 + 0.000073I)$
South Central	1529.1	−3.0857	$Q = 273.313 + (P - 10.299)(2.96201 + 0.000258I)$
Rocky Mountains	903.6	−2.7106	$Q/[I(P - 10.299)/26.09] = 0.00349 + 0.0021P$
Pacific Southwest	768.9	−2.5127	$Q/[I(P - 10.299)/20.45] = 0.01968 + 0.00042P$
Pacific Northwest (West)	2752.9	−4.5153	$Q/[I(P - 10.299)/20.74] = 0.03107 + 0.00111P$
Pacific Northwest (East)	551.2	−1.7226	$Q/[I(P - 10.299)/19.94] = 0.0350 + 0.000111P$

Note: FI = Forest industry; OP = other private forest landowners; Q = quantity harvested (MMcf); P = stumpage price (1967$/Mbf); and I = stumpage inventory (MMcf). Mbf = thousand board-feet; MMcf = million cubic feet.

[a] Derived demand equation: $Q = Q_d(P)$.

[b] Supply: FI and OP equations are unchanged from the base case. (See table 2 in this paper.)

Table 4. Differences Between 1977 Actual and Simulated Market Clearing Quantities and Prices

Region	Harvest levels				Stumpage price (1977$/Mbf) (actual–simulated)
	% changes by ownership			Total (in MMcf) (actual–simulated)	
	FI	OP	Public		
Southeast	−17	−5	+425	837.07–1,016.73	95.06–76.75
South Central	−7	−5	+183	1,387.40–1,541.10	104.89–83.54
Rocky Mountains	+25[a]	[a]	−19	861.94–832.93	30.41–50.67
Pacific Southwest	−4	−14	+8	711.65–717.53	66.71–39.71
Pacific Northwest (West)	−34	−74	+70	2,565.33–2,651.30	99.04–40.28
Pacific Northwest (East)	−3	−35	+6	511.01–516.89	56.96–38.72

Note: FI = Forest industry; OP = other private forest landowners. Mbf = thousand board-feet; MMcf = million cubic feet.
[a] In the Rocky Mountain region, FI and OP are combined.

nitude for these quantity and stumpage price shifts fit the consensus intuition of policy analysts who have considered these three regions. (See, for example, Adams and coauthors [1982]; Hyde [1981, 1983]; Walker [1974].)

The Rocky Mountain region displays the opposite price and quantity effects. Public harvests decrease substantially, private harvests increase sharply but not enough to offset the public decrease, and the regional stumpage price increases sharply. Table 5 shows the net welfare gains for both private and public producers. Private producers gain from both price and harvest increases; public producers gain by decreasing ineffective production activities. The directions of these Rocky Mountain region effects satisfy our impression of consensus intuition for all but that share of the local forest products industry that depends on public timber as a resource input. We expect that a more careful analysis would suggest even more substantial public harvest decreases in this region.

The aggregate (public and private) 1977 harvest shift is not large—only a 5.8 percent increase—but the regional price adjustments are meaningful. The price adjustments in each region are greater than we expect within a normal year. They are not as great, however, as the adjustments we have observed in the most recent ten years (a turbulent time for the industry), and they are also not as great as observed price variations within regions but across quality, species, access, and ownership classes at any moment in time. (Observed interregional variations exceeding several hundred percent are common.)

The general price decrease across the aggregate of all regions suggests that consumers (intermediate-good producers and final-good consumers) obtain the greatest gains from a public agency shift to market-sensitive harvest criteria. Price decreases should also cause some counterpart losses by timber producers, losses that would be further emphasized for private producers in all regions except the Rocky Mountains, because private producers also lose market shares to the public sector. These trends are exactly what table 5 shows. It should be recognized that consumer gains in table 5 refer to gains for consumers of timber produced in the identified region. The intermediate- and final-good consumers themselves may not live in the producing regions.

We expect large public-sector efficiency gains as a result of both the shifts in level of public harvest and the change to market-responsive public cost functions. Yet the ad hoc nature of our public cost estimates for both the base case and the simulated market solution yield less confidence in our estimates of public surplus gains. Nevertheless, our large estimates (greater than six times the actual 1977 public timber management expenditures) strongly suggest that there are great gains for the public treasury to be found in modified public timber management.[5]

In conclusion, the shifting of public timber management criteria to reflect market sensitivity produces a large aggregate social gain and a large gain for the public treasury. There are also underlying losses for many private timber producers and perhaps for both timber producers and consumers in one region,

Table 5. Efficiency and Distributive Gains from Fully-Market-Responsive Timber Harvest Criteria (millions of 1977 dollars)

Region	Consumers' surplus	Producers' surpluses			Net social welfare
		FI	OP	Public	
Southeast	40.2	−21.4	−57.5	179.8	151.1
South Central	45.3	−74.2	−76.5	128.2	12.8
Rocky Mountains	−30.9	45.8[a]	—[a]	41.0	55.9
Pacific Southwest	20.6	−69.5	−11.9	—[b]	—[c]
Pacific Northwest (West)	689.8	−398.3	−68.6	996.8	1,219.7
Pacific Northwest (East)	21.8	−12.2	−3.3	—[b]	—[c]
Total	786.8	−592.8	−217.8	—[c]	—[c]

Note: FI = Forest industry; OP = other private forest landowners. Dash = not available.

[a] In the Rocky Mountains region, FI and OP are combined.

[b] Ambiguous because the highly elastic public supply in this region prevents the model from finding a stable solution within a small number of iterations.

[c] Total not given because of uncertain public producers' surpluses in the Pacific Southwest and the Pacific Northwest (East).

the Rocky Mountains. Increasing timber harvests raises the question of associated environmental damage. Yet, in the more fragile forests of the Rocky
Mountains there is, no doubt, a substantial by-product gain in nonmarket forest
outputs as a result of following market timber criteria.

The results of our study refer to 1977, but they also suggest price- and
inventory-destabilizing effects and social welfare gains in the near term after
1977. Although these near-term adjustments are large, we expect that the
long-term price and inventory effects may be less important. Long-term market
pressures will deplete the public inventory surplus in a market-responsive case.
Long-term market and political pressures may force a gradual administrative
release of public inventories even if current public timber management criteria
remain in effect. Long-term welfare gains are less determinate. They will persist,
regardless of price and inventory levels, to the extent that the public sector
becomes responsive to production costs; that is, permitting marginal costs to
determine supply is less expensive than the current practice.

The dramatic market impacts suggested by our analysis argue for further
and more precise research on this topic. If further research results are consistent
with these impacts and nonmarket values do not weigh more heavily when
balanced against them, then these effects argue for a change from our current
market-constraining public management policies.[6]

Recreation Management

Supply is the focus of this section, which considers a nonmarket recreation
service provided on public forest lands. Nonmarket values usually suggest
concentration on demand estimation, and there is an extensive literature on
this topic. This means, however, that the supply side of efficient resource
allocation is somewhat overlooked. Nevertheless, we can show that supply data
are readily available in the form of public agency budget and expenditure
records, and we can reasonably argue that land managers possess the insight
to make judgments more easily and confidently about supply than about
demand. Furthermore, accurate supply information alone encourages least-cost
decisions and can often prevent egregious errors in resource allocation.

The objective of this section is to consider recreation resource allocation;
the particular resource under consideration is U.S. Forest Service campgrounds—a type of recreational resource that is also common to various federal
and state agencies and even some private operations. The magnitude of public
agency recreation budgets and levels of visitor use alone suggest the significance
of possible errors in the allocation of recreational resources and the potential
importance of improving our understanding of it. In Fiscal Year 1984 the
Forest Service spent $99 million providing 229 million visitor-days of outdoor
recreation (G. Elsner, recreation economist, U.S. Forest Service, Washington

office, personal communication, February 1985). The National Park Service spent an additional $104.9 million providing another 108 million visitor-days (U.S. Department of the Interior, National Park Service, 1984, 1986). Clearly, errors in recreation resource allocation can have large social impacts; thus, we can make some contribution to more efficient allocation even if our insights are limited to suggesting small adjustments in these budgets or their distributions.

We approach the question of recreation resource allocation through a case study of national forest campgrounds in the Seeley–Swan Valley (SSV) in western Montana. The valley, which is long and narrow, is bordered by the Mission and Swan Mountains to the west and east, respectively. The Swan River flows the length of the valley and drains the several lakes that are the main attraction at four relatively similar Forest Service campgrounds.[7] The valley is a popular recreation spot within easy access of 42 percent of Montana's population. It is relatively unknown, however, to the out-of-state campers travelling the interstate highways just north and south of it en route to Glacier and Yellowstone National Parks. In sum, the SSV might be characterized as providing an important though not unique recreation opportunity for a regional market.

The resource allocation problem confronting SSV managers is illustrative of the general problem facing on-the-ground public agency managers. The important constraints involved in the problem are the physical boundaries of agency jurisdiction rather than the greater market area and the agency's budget appropriations. The timing perspectives we adopt for the analysis are those of the agency's local recreation manager in the off-season—for example, December—as he or she plans the next year's operation from the base of whatever campground facilities are already in place. Expansion is unlikely under current public agency budgets, but altering use patterns, some reduction in facilities, various maintenance activities, and even the possibility of contracting for private operation of the campgrounds are all options for the manager at this time.

We approach the allocation problem through marginal analysis in which incremental expenditure options define supply schedules for the four SSV campgrounds and aggregate consumer preferences define a single demand schedule for all four. The following subsections first construct the supply schedules and then present results and conclusions. The Forest Service recreation management decision as it is made under the present system begins with an externally established arbitrary fee for campsite use. (Three of the SSV campgrounds charge $4 per campsite per night; the fourth, which adds flush toilets, charges $5 per night.) The recreation manager has discretion only with respect to cost allocation. The contrast between the current level of campground operation under these arbitrary prices and the optimal improvement that might be made after considering our supply information and the implicit demand schedule is the main feature of our conclusions.

Supply Schedule Construction

The annual planning perspective of this analysis suggests that the existing level of permanent Forest Service personnel and the four existing campgrounds, each with its current number of campsites, are sunk costs. It is not within the discretion of recreation managers to alter either of these elements in the coming year. Discretion pertains only to the fixed costs of opening each campground for annual operation and to the variable costs associated with the level of campground use. Fixed costs may be constant across campgrounds, or they may vary according to the number of existing and identifiable campsites within each campground. Variable costs are functions of the number of recreation-visitor-days (RVD, where one RVD equals one camper for one twelve-hour period) of campground use.

The data for our analysis were supplied by the Forest Service's Program Accounting and Management Attainment Reporting System (PAMARS). These data for the four specific SSV campgrounds have the advantage of reflecting actual 1984 expenditures rather than *ex ante* budget allocations. The local recreation staff helped us interpret the PAMARS data so that we could divide them into twelve recreation activity groups from which we developed twelve underlying cost functions. Together these form the total cost function for each campground.

Tables 6 and 7 summarize the cost functions for each activity. The campground is the basic observational unit. Table 6 identifies the fixed cost and the variable cost coefficients associated with each activity in the campground. (The regressions repress the intercept terms.) Fixed costs can be functions of either opening the campground for annual operation or of the number of campsites per campground. (All of the campsites in each campground are independent and well marked.) The variable costs are functions of the 1984 use levels. Table 6 also reports three tests for statistical fit. The F tests all reject at the 10 percent level the hypothesis that all equation coefficients are equal to zero.

The small sample size requires that we construct, rather than statistically estimate, four underlying total cost functions: utilities, replacement required because of physical deterioration, fee management, and vehicle use. Table 7 shows these constructed costs. The assistance provided to us by the local SSV recreation staff improves our confidence in these costs.

Opening the Campground Two activities listed in table 6 involve the only single and permanent costs associated with opening the campground: land use and utilities. The cost of assigning the lands in these campgrounds to recreation rather than to some other use is an implicit cost equal to the net value of the lands' forgone next-best use. Timber and water production are the likely alternatives for any of the SSV campgrounds. But these campgrounds are all in fragile riparian areas where Forest Service policy seeks to protect water

Table 6. Recreation Activity Total Cost Functions for Four Seeley–Swan Valley Campgrounds, 1984

Activity	Fixed costs		Variable costs per RVD	R^2	F
	Camp-ground	Campsite			
Land use	0	—	—		
Electricity	a	—	—		
Toilet pumping	—	5.1387	0.00135	.999	890.4**
		(15.099)**	(1.036)		
Garbage collection	—	3.8005	0.0418	.919	11.31*
		(0.474)	(1.362)		
Maintenance—user-induced	—	2.4757	0.0103	.997	281.2**
		(4.409)**	(4.792)**		
Prevention of physical deterioration					
Labor	—	− 19.9448	0.3191	.985	67.07**
		(− 1.346)	(5.624)**		
Facility repair	—	2.578	0.0794	.989	86.55**
		(0.558)	(4.487)**		
Replacement	a	—	—		
Compliance checks	—	3.9364	0.08268	.993	145.1**
		(0.020)	(5.595)**		
Fee management	—	—	a		
Law enforcement	—	29.9695	0.0197	.959	23.63*
		(2.254)	(0.387)		
Vehicle use	a	—	a		
Overhead	0	0	0		

Note: RVD = recreation-visitor-days. Numbers in parentheses are t-statistics. Dash = not applicable.

*Significant at the 0.10 level.
**Significant at the 0.05 level.
[a]See table 7 in this paper for constructed costs.

quality by requiring longer timber rotations, and even then harvests are re-stricted within some distance of streams and lakesides. This suggests that harvests would seldom, if ever, occur on the sixty-eight acres in the four campgrounds, so the expected timber value, which is also the campground opportunity cost, is effectively zero. The Forest Service managers judged that SSV campground use does not endanger water quality, so this potential op-portunity cost is also zero.

The local electric utility sends a monthly bill for each campground, but each bill is close to the minimum charged to provide service to that facility. That is, once the electricity is turned on and the campground is open for

Table 7. Constructed Costs for Four Recreation Activities at Seeley–Swan Valley Campgrounds, 1984
(in 1984 dollars)

	Campground			
Activity	Lake Alva	Big Larch	River Point	Seeley Lake
Utilities,				
per campground	120.00	137.60	103.57	289.00
Replacement due to				
physical deterioration,				
per campground	5,300.60	9,018.16	3,397.64	7,692.28
Fee management per RVD	0.0791	0.0709	0.0715	0.0309
Vehicle use				
Per campground	1,668.62		2,669.07	
Per RVD	0.0869	0.0210	0.0267	0.0142

Note: RVD = recreation-visitor-day.

visitor use, the level of use is seldom sufficient to raise the electricity bill above the minimum fee. Therefore, this fee is equivalent to a one-time-only annual fee. (There are no significant statistical relationships between the electric bill and either RVDs of use or numbers of campsites.)

Complex Cost Functions   Estimating the cost functions for the remaining accounting activities is more complex, since most of them have both fixed-cost and variable-cost components and two (vehicle use and overhead) are joint costs. The Forest Service contracts with private firms to perform two of the activities, which means that the cost reporting for these two activities is different.

Recreation managers locate toilets and garbage dumpsters near campsite groups within each campground and contract both toilet pumping and garbage collection to private firms. The fee for these activities is a function of the number of trips the contractor makes. A contractor makes a trip, emptying every toilet or dumpster within a campground, each time one toilet or dumpster within the campground becomes full. Therefore, it is not surprising that costs for toilet pumping and for garbage collection are functions of both the fixed number of campsites (which relates to the number of toilets and dumpsters) and the variable number of RVDs of use (which dictates the number of trips) within each campground.

Maintenance includes two accounting activities: (1) user-induced maintenance and cleanup and (2) repairs required as a result of normal physical deterioration. The user-induced maintenance and cleanup cost equals the time spent on maintenance multiplied by the Forest Service wage ($0.857 per hour) for summer volunteers. User-induced maintenance is distributed across campgrounds according to the number of campsites and the level of use within each campground. (Some capital facilities require periodic maintenance, even if they are only used once, plus additional maintenance depending on their use levels.)

Whether maintenance to prevent normal physical deterioration is a fixed or variable cost could be the subject of reasonable debate. Forest Service policy limits allowable physical deterioration, requiring that all capital be maintained in high-quality condition. Therefore, local recreation managers do not have the option of permitting even an unused campground to deteriorate, and the costs that must be incurred to prevent normal deterioration are fixed costs. Conversely, our regressions show highly significant relationships between these costs and campground use level. This significance might reflect a coincidental relationship between (1) normal deterioration and fixed campground size and (2) the use level. Therefore, the relationship could be consistent with the fixed-cost argument. Table 6 records the statistical results of our analysis. Our later work treats maintenance to prevent deterioration as a fixed cost, but it also examines the change in optimal resource allocation when deterioration is treated as also having a variable component. The allocation decision for this cost is important because the variable component is 60 percent of total variable costs.

Maintenance to protect against normal physical deterioration has both labor and capital cost components, and the capital component further divides into facility repair and replacement costs. We estimate the labor costs as the residual Forest Service labor expenditures after deducting for labor used in all other recreation activities. The statistical distribution of these residuals across campgrounds is a function of both general campsite aging (or the number of campsites) and the campground use level. The statistical distribution of facility repair costs is similar.

Replacement cost is the value of existing capital in place depreciated at the Forest Service rate of 4 percent per annum. The estimates in table 6 derive from the capital inventory for each campground and the Forest Service replacement cost for each inventory item. Variations in terrain and access at the different campgrounds affect large-cost items such as roads and water systems, thereby preventing any obvious relationship between replacement costs and numbers of campsites.

Three camper management activities are included in the analysis: (1) compliance checks, (2) fee management, and (3) law enforcement. Compliance checks involve inspection of the campground to ensure that campers use only assigned campsites and are otherwise in compliance with the regulations for campground use. The cost of this activity is equal to the time it takes multiplied by the wage for (nonvolunteer) seasonal Forest Service employees ($7.23 per hour). The time spent is, in turn, a function of campground size and camper use.

(That is, larger campgrounds take longer to cover, and more campers mean more occupied campsites to check.)

Campers deposit their user fees in envelopes dispensed by the Forest Service. The envelopes also contain survey questions about the users. Fee management involves emptying these envelopes and sorting their user data. Clearly,

$$\text{fee management cost/RVD} = (\text{cost/envelope}) / (\text{RVDs/party})$$

where there is one envelope but perhaps many RVDs per party. The cost per envelope is the $7.23 hourly wage for regular hourly employees times the five minutes spent on each envelope. The number of RVDs per party is one of the survey questions on each envelope. Seeley Lake campground, perhaps because of its flush toilets, attracts more users and for longer periods. Therefore, it accumulates more RVDs per party but lower incremental fee management costs per RVD than the other three campgrounds.

Law enforcement, which is conducted by other Forest Service personnel and charged against the recreation budget, is a high-cost item. The Forest Service recreation staff could not provide good arguments for the determining factors behind the allocation of the cost to particular campgrounds. Our weak statistical relationship suggests that the number of sites, each of which must be patrolled whenever the campground is open, and the use level of each may affect costs.

Finally, the two costs, vehicle use and overhead, are joint to many of these activities. The Lolo National Forest, in which the Seeley–Swan Valley is located, charges its recreation budget a total vehicle cost for use in association with all recreation activities except the contracted activities. The recreation staff recommends a division of this cost across all four campgrounds in such a way that a larger share accrues to Lake Alva, the most distant from the other three campgrounds. We somewhat arbitrarily divide the share for each campground into fixed and variable components according to the fixed–variable share of the sum of all noncontracted activities and further divide each variable share by the use level for that campground. The fixed-cost shares for the three campgrounds that are adjacent to one another are truly a joint cost because Forest Service personnel normally travel to all three of these in the same trip.

The Lolo National Forest supervisor's office and the Forest Service regional office charge overhead to the ranger district as a share of the total field expenses for personnel and systems support. We argue that this charge is a sunk cost that is irrelevant for our objective of examining recreational operations for a single season. The basis for this argument is that, although the supervisor's office and the regional office do alter the magnitude of their staffing, and, therefore, the magnitude of their total overhead charges, such alterations only occur in response to the demands of a longer term than one season.

Empirical Results and Conclusions

We can combine our supply information with a knowledge of consumer demand in order to learn about the socially optimal price and use levels for the campgrounds and to compare the social welfare and recreation management differences between these optimal levels and the actual 1984 levels.

Demand information originates from the user survey data (postal ZIP code of user origin) on the reverse side of the fee envelopes. The specification for the inverse travel cost demand function for the aggregate of all four campgrounds is as follows:

$$Q = e^{\alpha}(TC + F)^{-\beta}$$

where Q = RVDs per 1,000 population in the origin zone
TC = implicit travel cost per RVD (composed of the round-trip mileage cost and the values of both travel time and onsite time)
F = implicit campsite fee per RVD
Q_c = actual 1984 RVDs (37,543)
α, β = parameters

The α is normalized such that it defines a function passing through observed values for Q and $TC + F$ for the 1984 season. The estimated values for α and β are 40.9073 and 6.5136, respectively. The standard error on β is 0.7350.[8]

This demand function is conceptually identical to that developed by Wilman (1980), and its results fall within the range of the Sorg and Loomis (1984) survey. Furthermore, the statistical evidence strongly supports our hypothesis that the four campgrounds are homogeneous in their demands. That is, the estimated parameters for the four independent campground demand functions all fall within one standard error of each other and within one standard error of the estimated parameter for the aggregate demand function.

Our first results derive from the joint allocation of the vehicle use cost. They imply a sequence in the annual opening of the campgrounds: if it is optimal to open only one, then Lake Alva is the least expensive to operate. If opening more than one campground is optimal, then two or more of the other three are least expensive. This issue is not of major concern for SSV recreation, however, because the annual use levels of the campgrounds under any reasonable pricing scenario recommend opening all four.

The other results derive from setting the marginal cost equal to the implicit campground user fee. As shown in table 8, the 1984 implicit fee per RVD was $0.81 at Lake Alva and in the range of $0.71 to $0.77 at the other three campgrounds. The aggregate use level was 37,543 RVDs, and the net economic value, including consumers' and producers' surpluses, was $609,820. Setting the implicit fee equal to the marginal cost changes the use level to somewhere

Table 8. Comparison of Actual and Optimal Seeley–Swan Valley (SSV) Recreation Use Levels

Fee level	Lake Alva	Three other SSV campgrounds	Use level (RVDs)	Net economic value
Implicit	$0.81	$0.71–$0.77	37,543	$609,820
Low optimal	$0.32	$0.25	38,695	$628,063
High optimal	$0.72	$0.65	37,673	$612,829

Note: RVD = recreation-visitor-day.

in the range of 37,673 to 38,695 RVDs and the net economic value to somewhere in the range of $612,829 to $628,063. (The treatment of maintenance to prevent deterioration as either a fixed or variable cost determines these ranges.)

It is clear that there is some improvement in social welfare from decreasing the fee and increasing use until the fee equals the marginal cost. It is also clear, however, that a large percentage decrease in the fee has only a small effect on the use rate and on social welfare because the fee is only a small share of the implicit travel cost (approximately $106 per RVD). Furthermore, whereas setting the fee equal to the marginal cost has a small effect on social welfare, it has a large negative effect on Forest Service recreation receipts. It may be a wise public agency administrative and political choice to set the fee above marginal cost and thereby obtain receipts that are much closer to the total costs of the recreation operation, making SSV recreation much less subject to the whims of the federal budget process. (See Freeman and Norris [in this volume] for a discussion of this point.)

Regardless of the administrative and political advisability of the choice of fee level, the marginal cost is constant per RVD, and changes in the demand price have little impact on the optimal use level. (The supply function is elastic and the demand function inelastic.) Furthermore, the optimal and actual use levels are similar. The case of the SSV campgrounds is certainly one of those satisfying instances in which analysis suggests that society is doing well by relying on the manager's intuition.

This first-generation research shows the usefulness of recreation supply-and-demand analysis and the sources of both data and managerial skills regarding supply. It also raises interesting questions about optimal pricing schemes when we generalize the results of this analysis to other recreational opportunities. The SSV facilities are similar, and the supply-and-demand elasticities are uniform across them. Generalizing to other recreation opportunities ranging from undeveloped wilderness camping to resort communities changes the uniformity of elasticities. Such a range of opportunities can occur even within the management responsibility of one Forest Service ranger district. Therefore, this new pricing issue can be a serious problem even for local managers.

Let us consider multiple classes of recreation opportunities and, for example, a management decision to constrain operations in such a way that revenues exceed some share of total costs. (This constraint is similar to the apparent intent of management at the SSV campgrounds.) Efficiency criteria and social welfare maximization remain the guiding characteristics within the constraint. The optimal pricing scheme—and a measure of its variation from the current price policy—becomes an interesting question for the next generation of public recreation policy analysis.[9]

Conclusions and General Implications

Public land management is probably the most thoroughly reviewed and analyzed topic in all of the literature in the field of natural resource management. This paper makes no attempt to repeat this literature or to be comprehensive in its own review. Rather, it samples two unresolved and contemporary public land management issues with the intent of raising the quality of analysis and the level of discourse related to them.

The first issue involves timber, a private good produced on the public lands. The second issue has to do with recreation, a service with public good aspects that is often produced on many of the same public lands. Timber and recreation are probably the largest-volume and the most highly valued of all the multiple uses of the public forest lands. The market fully values and freely exchanges timber; yet the public forestry agencies, in general, prefer nonmarket criteria for timber harvest decisions. We examine the opportunity costs of the departure of such agencies from market criteria with respect to both sale prices and production costs. Our recreation analysis focuses on the Seeley–Swan Valley in western Montana. It searches for an optimal demand–supply equilibrium, but its main feature is the estimation of a local recreation supply function, something that is overlooked by most of the recreation economics literature but that is nonetheless a key item for on-the-ground recreation managers. Supply may be particularly important because it is the only factor that managers can alter.

The public lands produce 22 percent of the annual softwood timber harvests in the United States. The timber management objective of the agencies that oversee such lands resembles the production of an even flow of annual sawtimber harvests regardless of local market criteria. Special interest groups widely discuss the resulting departures from market results, but there has been no general analysis and there is no consensus of opinion. Our approach to this problem begins with the simulation of market-responsive public supply functions for six regions of the country. We then introduce these functions into the Adams–Haynes aggregate model of the timber economy, which permits us to contrast regional prices, quantities, harvest levels, and consumers' and producers' surpluses occurring under conditions of either current or market-sensitive public timber management criteria.

Our analysis is impressionistic and should not be taken as conclusive. It represents a one-year harvest shift and incorporates no dynamic effects, no local variations (which might dominate political decisions regarding regional solutions), and no adjustments for nonmarket values that might compete with timber production. We mean to leave no implication of precision.

Nevertheless, our results are truly large and emphatic. They suggest substantial harvest increases on the public lands for all regions except the Rocky Mountains. These increases would cause stumpage price decreases, large consumer gains, large gains for the public treasury, and some offsetting harvest and profit decreases for private timberland owners (see table 5). The welfare shifts are large enough, if accurate, to dominate the welfare impacts of most harvest regulations of private forest lands. Furthermore, while these welfare gains are due to timber harvest increases, they will be associated with local harvest reallocations which often yield environmental gains. Although we make no claim for accuracy in the magnitude of our results, we do believe that they emphasize the importance of this policy issue. The departure of public timber management from the use of market harvest criteria deserves further serious inquiry.

The public lands also provide more than 300 million annual visitor-days of outdoor recreation at a cost that exceeded $1.1 billion in 1984. As with timber, special interest groups question this allocation and this budget level, but there is little general analysis and no consensus of opinion. Our approach to the recreation allocation question contrasts with our timber analysis in that it focuses on one particular valley and one level of recreation quality in its search for optimality.

The recreation demand analysis presented here uses the common travel cost method. Our supply analysis depends on cost functions constructed from local Forest Service data with the considerable assistance of the local managers themselves. The results are satisfying for both managers and analysts (see table 8). They show that equilibrium price and quantity levels for the four campgrounds are close to the actual price and quantity levels, and the small deviation of the actual from the optimal in this case makes good budgetary sense. In other words, the SSV recreation managers perform well by social criteria. It may also be that recreation supply has not been a critical analytical problem and that the absence of an abundant literature reflects this.

Of course, it is impossible to generalize from the small sample of four campgrounds in the Seeley–Swan Valley, but our analysis should leave both managers and analysts with a feeling of satisfaction. We can only assume that public agency recreation management in the SSV is comparable to public agency management elsewhere. Therefore, relatively optimal allocations here suggest that departures from optimality may not be either a frequent or a great problem elsewhere. Our analysis begs replication for the purpose of testing this hypothesis.

Several messages emerge from these two illustrative inquiries. First, there is much to be gained from pragmatic analyses such as these. Both the timber and the recreation cases display an easier understanding of critical issues than many would expect prior to performing the analysis. The data are available and public agency managers possess the critical skills necessary to perform such analyses for other public land issues and at other locations.

The timber and recreation inquiries described in this paper encourage the hypothesis that budget restrictions keep management lean—and close to efficient—whereas market-valued goods and an inflow of cash encourage agency growth and deviations from social welfare. Does public forest land agency provision of other nonmarket or public goods (for example, other forms of recreation) support this hypothesis by displaying approximate efficiency? Does the provision of other market-valued goods (for instance, range or minerals) support the hypothesis by displaying sharp deviations from efficiency and substantial social welfare costs?

Finally, the SSV recreation case raises potentially important questions about public good pricing under budgetary constraint. For example, when does the constraint begin to push allocation sharply apart from social optimality? (Apparently the constraint has little impact on optimality in SSV.) What does the budget constraint do to optimal allocation across several campgrounds of different quality (and therefore of different demand elasticities) within the same management jurisdiction? This is a common problem for many Forest Service ranger districts and for most national forests.

Of course, these do not exhaust the set of issues relevant to public forest land management. It is our belief that these too can be addressed, at a pragmatic level, with analysis and data comparable to what we used in these case studies. We mention them in closing, because they bring us back to where we began— to the research contributions and leadership of John Krutilla.

Public forest land management will need to consider:

1. the opportunity costs of endangered species management;
2. the welfare costs of "community stability" goals in relation to its *actual* employment-generating effects;
3. the income distributional effects of outdoor recreation policy;
4. the role of nonuse values in allocation decisions for public lands—both unique and nonunique resources; and
5. the implications of trade policy—in particular, public land log export restrictions for management of public forests.

We can begin to address each of these. Indeed, Krutilla has already pointed us in the right direction on the first four issues—with Krutilla and Fisher

(1975) for items 1 and 4, Haveman and Krutilla (1968) for the second, and Krutilla and Eckstein (1958) for the third.

In conclusion, neither of our analyses provides universal evidence of a high degree of confidence regarding the welfare impacts of any public land management policy. Each, however, provides an emphatic statement about an important aspect of public land management, and their results show why resource analysts have put great effort into understanding public land policy. It is a broad and critical issue, one that begs further inquiry into its many and diverse policy aspects.

NOTES

1. That is, our analysis compares current expenditures with current receipts. Yet the correct comparison is between current expenditures and the discounted future receipts that managers anticipate those expenditures will create. If planned harvest volumes remain constant over time and output prices and factor costs retain their current relative relationships, then current receipts are a reasonable proxy for discounted future receipts only if capital has a zero opportunity cost. Our true 1977 cost estimate is conservative if the actual cost of capital exceeds zero.

2. Adams and Haynes include more sophisticated intermediate-product demand functions for these two regions. To a large extent, intermediate-product demands in these regions are originally satisfied by stumpage supplies from the other six regions.

3. David H. Jackson, associate professor of forest economics, University of Montana (personal communication, 1985) found a difference of $12 per thousand board feet between state and private stumpage in Montana. The federal–private difference is probably greater. Furthermore, it is also our observation that federal accounting procedure's overestimating of the public–private difference only causes underestimation of the anticipated social welfare losses due to public agency constraints on market timber criteria—and thereby strengthens our eventual conclusions.

4. Darius Adams and Joanna Kincaid assisted us in our table 3 revision of the AH model. The interpretation of results remains our own.

5. Public-sector gains are measured as follows: 1977 actual = price (table 4) times harvest quantity (table 2) minus costs (table 2); simulated = the area below equilibrium price (table 4) and above the supply schedule (table 3).

The 1977 actual net public timber revenues can be negative. The public procedures surplus calculated in the simulated case is due to the fixity of capital in cases in which capital is the standing economic inventory. Table 5 shows the difference between the 1977 actual net gains and the simulated market-sensitive net gains.

6. John V. Krutilla's currently active research (with Michael D. Bowes) investigates the effects of nonmarket forest values on optimal economic choices for timber management (Bowes and Krutilla, in press).

7. All four campgrounds fall within the bounds of the Seeley Lake Ranger District. The district itself extends beyond the Seeley–Swan Valley and includes five other similar campgrounds in neighboring supply sheds.

8. The specification of the demand function is log linear generalized least squares. Unequal zonal populations cause heteroskedasticity; therefore, we weight the regression by the square root of zonal population. This is equivalent to generalized least squares and, as such, is unbiased and efficient. Travel time is valued at 35 percent of the average wage rate; on-site time is 100 percent of the average wage rate. (These valuations are consistent with Wilman's [1980] theoretical recommendations.) Travel is at 50 miles per hour, and the average wage rate within each zone of origin for recreationists is from the Bureau of the Census's *Local-Personal Income 1977–82* (U.S. Department of Commerce, Bureau of the Census, 1984). We derived α because otherwise the regressions produced statistically insignificant estimates of it and also yielded meaningless consumers' surplus values.

9. V. Kerry Smith clarified this question for us.

REFERENCES

Adams, D. M., and R. W. Haynes. 1981. *The 1980 Softwood Timber Assessment Market Model: Structure, Projections and Policy Simulation.* Forest Science Monograph 22. Supplement to *Forest Science* vol. 26, no. 3 (Washington, D.C., Society of American Foresters).

Adams, D. M., R. W. Haynes, G. F. Dutrow, R. L. Barber, and V. M. Vasievich. 1982. "Private Investment in Forest Management and the Long-term Supply of Timber," *American Journal of Agricultural Economics* vol. 64, no. 2, pp. 232–241.

Bowes, M. D., and J. V. Krutilla. In press. *Multiple-Use Management: The Economics of Public Forestlands* (Washington, D.C., Resources for the Future).

Haveman, R. H., and J. V. Krutilla, with R. M. Steinberg. 1968. *Unemployment, Idle Capacity, and the Evaluation of Public Expenditures* (Baltimore, Md., Johns Hopkins Press for Resources for the Future).

Hyde, W. F. 1981. "Timber Economics in the Rockies," *Land Economics* vol. 57, no. 4, pp. 630–639.

————. 1983. "Development vs. Preservation in Public Resource Management," *Journal of Environmental Management* vol. 16, no. 3, pp. 347–355.

Krutilla, J. V., and O. Eckstein. 1958. *Multiple Purpose River Development: Studies in Applied Economic Analysis* (Baltimore, Md., Johns Hopkins Press for Resources for the Future).

Krutilla, J. V., and A. C. Fisher. 1975. *The Economics of Natural Environments: Studies in the Valuation of Commodity and Amenity Resources* (Baltimore, Md., Johns Hopkins University Press for Resources for the Future).

Krutilla, J. V., and A. C. Fisher, with R. E. Rice. 1978. *Economic and Fiscal Impacts of Coal Development: Northern Great Plains* (Baltimore, Md., Johns Hopkins University Press for Resources for the Future).

Sorg, C. F., and J. B. Loomis. 1984. "Empirical Estimates of Amenity Forest Values: A Comparative Review." U.S. Department of Agriculture—Forest Service General Technical Report RM-107 (Fort Collins, Colo., Rocky Mountain Forest and Range Experiment Station).

U.S. Department of Agriculture. 1969–1983. *Agricultural Statistics* (Washington, D.C., Government Printing Office).

U.S. Department of Agriculture, Forest Service. 1978. "Forest Statistics of the U.S., 1977." Review draft. Washington, D.C.

————. 1982. *An Analysis of the Timber Situation in the United States 1952–2030.* Forest Resource Report No. 23 (Washington, D.C., Government Printing Office).

U.S. Department of Commerce, Bureau of the Census. 1984. *Local-Personal Income 1977–82* (Washington, D.C., Government Printing Office).

U.S. Department of the Interior, National Park Service. 1984. "133 Report on Budget Execution" (Washington, D.C., U.S. Department of the Interior, National Park Service, September 30).

————. 1986. *National Park Statistical Abstract, 1985* (Denver, Colo., National Park Service, Denver Service Center).

Walker, J. D. 1974. "Timber Management Planning." Mimeo. Western Timber Association, Portland, Oregon, August.

Wilman, E. A. 1980. "The Value of Time in Recreational Benefit Studies," *Journal of Environmental Economics and Management* vol. 7, no. 3, pp. 272–286.

PART 3

RESOURCE POLICIES AND THE PRACTICE OF APPLIED WELFARE ECONOMICS

MODELING RECREATION DEMANDS FOR PUBLIC LAND MANAGEMENT

Elizabeth A. Wilman

The need to make rational decisions with regard to the allocation of natural resources increases as demand for the products of the public lands—wood, water, forage, wildlife habitat, and recreational opportunities—grows while the supply of natural resources available to meet these demands stays relatively stable. The pioneering work of John Krutilla has been influential in showing the relevance of the analytical, management, and policy tools of efficiency economics to this problem (Bowes and Krutilla, in press; Fisher and Krutilla, 1972; Haigh and Krutilla, 1980; Krutilla, 1966; Krutilla and Fisher, 1975; Krutilla and Knetsch, 1970; Krutilla, Bowes, and Wilman, 1983).

In an unconstrained world, the objective of multiple use management in terms of economic efficiency is the maximization of the present value of net benefits to society, with society being defined as the political jurisdiction that exercises control over the resources. Obtaining this maximum involves manipulating the mix of uses of the vegetative, land, and water resources of public lands in order to produce a desired combination of timber, forage for cattle, water supply, wildlife habitat, and recreational opportunities. In a more constrained world the goal might be more limited to simply improving the efficiency with which resources are allocated, in which case the objective would just be for a given management action to produce a positive present value of net benefits.

Unfortunately, some of the outputs of public lands either have no market prices or have prices that are unrelated to their economic values. This lack of market pricing may stem from historical or legal causes, or there may be technical reasons related to the degree to which it is possible to exclude nonpaying users. In any case, the lack of market prices means that it is necessary to determine the demands for these goods indirectly in order to measure the net benefits obtainable from various levels of these outputs. This paper focuses on one particular category of outputs—recreation opportunities. Describing the recreation opportunities offered by public lands requires descriptions of both the quantity and quality dimensions of the services they provide. However, with the exception of explicit restrictions (or removal of restrictions) on the level of use, most management actions are best treated as affecting the quality rather than the quantity of these services. This means that concern here is with the effect that public land managers have on the quality of recreation opportunities and on the present value of the resulting net benefits. John Krutilla's work has been pathbreaking in recognizing and analyzing the significance of people's willingness to pay for quality in relation to natural environments and public land management (Fisher and Krutilla, 1972; Krutilla, 1967 [reprinted in the appendix in this volume]; Krutilla and Fisher, 1975). Although Krutilla's work in this area has extended beyond on-site use, that is the focus of this paper, which discusses methods of estimating the benefits of quality changes to on-site users.

With the quality dimension as the focus of interest then, it is necessary to measure objectively the characteristics or attributes of recreation sites. These may include vegetative, landform, and water features and may also include facilities that are available at developed sites. At least some subset of the characteristics must be manageable. As long as consumers of recreation opportunities would be willing to pay more for an opportunity with more of the characteristic than for one with less of it, the difference in willingness to pay, net of cost, will measure the benefit that a consumer would obtain from an increase in the characteristic.

There are in general two ways by which economists have tried to measure the benefits from improvements in the quality of recreational opportunities. The first approach, which will not be further explored here, is the contingent valuation method. It uses direct questioning to obtain benefit estimates. This approach was originally formulated by Davis (1963). It has also been used by Cicchetti and Smith (1973), by Hammack and Brown (1974), and, more recently, by Mitchell and Carson (1981); Brookshire and coauthors (1982); Bishop, Heberlein, and Kealy (1983); and Brookshire, Eubanks, and Randall (1983).

The second approach uses information on the actual behavior of recreationists to infer their demands and the benefits they derive. The observations on behavior are usually cross-sectional; a data set would typically contain information on recreationists from a number of different places and information on the sites they visited. To interpret this multiple origin–multiple site data set, some model of consumer site choice is needed. If the quality characteristics of recreation sites are important, the site choice must be described in terms of these characteristics. A number of models have been used to explore consumer recreation site choices. Which of them is best for any particular case depends on which model best matches the data generated by that case. The sections that follow discuss and evaluate the two main classes of models—hedonic and multiple-site travel-cost—that have been used for these purposes. Stochastic elements with respect to these models are then discussed, after which a simple repackaging model is introduced as an alternative to the hedonic and travel-cost models. The possibility of changing visit lengths is then proposed. The final section summarizes the pros and cons of the three approaches and offers recommendations for further research.

The Hedonic Model

The hedonic model is largely the result of the work of Rosen (1974). While earlier contributions by Court (1941), Houthakker (1952), Tinbergen (1956), and Muth (1966) provide much of the background, Rosen formulated a spatial equilibrium model of product differentiation for market goods in which price differentials among goods with different amounts of a characteristic constitute information on the implicit, or hedonic, price of the characteristic. These hedonic prices are equilibrium prices. The model was used to recover estimates of the structural parameters of the marginal willingness to pay and marginal offer function. While these estimates are by no means as easy to obtain as Rosen implies, once they are obtained it should be possible to measure the benefits of increasing the level of a characteristic available in a given good.

Rather than presenting the full detail of the hedonic model, this section considers three crucial assumptions of the model that are of particular relevance in its application to recreation site choice. They are (1) the assumption precluding repackaging; (2) the requirement that if several quantity units are purchased, the same quality choices are made for each unit; and (3) the assumption that variation in cost conditions will identify a compensated demand curve.

No Repackaging

Let us first consider the no-repackaging assumption used by Rosen, which states that neither buyers nor sellers can (or find it economical to) repackage existing products. Rosen (1974, p. 38) cites two cases as examples of this assumption: "Two 6-foot cars are not equivalent to one 12 feet in length, since they cannot be driven simultaneously (case [1]), while a 12-foot car for half a year and a 6-foot car for the other half is not the same as 9 feet year round (case [2])."

The first case can be taken as a consumption constraint. In the recreation model the equivalent assumption is that only one site can be used per quantity unit (that is, per day or per visit). The second case essentially says that the simple repackaging model of Fisher and Shell (1971) and Muellbauer (1974) does not apply; that is, linear additivity of characteristic levels across quantity units does not apply. This is probably true for the site characteristics that are important for most types of recreation, although there may be exceptions in cases such as hunting and fishing. If the site characteristics include the probability of bagging an animal within a given time period (one day or one visit), then it may be that what is important to the hunter is the overall probability of bagging game (or the expected number of animals bagged) per season. A later section ("A Simple Repackaging Model") of this paper discusses a model for hunting that incorporates the simple repackaging concept.

Multiple Quantity Units

The second assumption of the hedonic model involves its extension to multiple quantity units. The hedonic model requires that the same choice in terms of quality be made each time a quantity unit is purchased. In the recreation case this means that, assuming days and visits can be used interchangeably as quantity units, all of the visits by one individual must be to the same site. But since the individual recreationist often uses many quantity units, this assumption is quite restrictive. Since the supply function for a characteristic upon which the individual must make a choice is the same for each visit, the requirement of same site choice under this assumption implies that the marginal willingness to pay for the characteristic must be the same across each visit and additive over visits.

In fact, Rosen (1974) carries the restriction one step further. Not only is the marginal willingness to pay for the characteristic constant and additive across quantity units, but so is the total willingness to pay for the bundle of characteristics.[1] If, in addition, the total willingness to pay for any quantity unit is identical to the total willingness to pay for the bundle of characteristics, a perfectly elastic demand curve for quantity units is implied. But the latter assumption does not necessarily hold in the recreation case, where the demand curve for visits is usually thought to be downward sloping, and the number of visits is determined by the equality of the marginal willingness to pay for a visit with the marginal cost of a visit. If this is the case, the marginal cost change for the characteristic, which also changes the marginal cost of a visit, would result in a change in the visit level. This would mean that, in addition to a demand curve for characteristics (which assumes a constant visit level), a demand curve for visits would have to be estimated in order to assess the benefit from a change in the supply of a characteristic.

On the other hand, a perfectly elastic demand curve for visits means that marginal willingness to pay and marginal cost only determine whether or not *any* visits will occur, not the level of visitation. Indeed, the latter must be determined by something outside this version of the model—for example, a fixed season length, or a time constraint that prevented more than a certain number of visits (if the amount of time required for a visit was large enough, no one might be able to take more than one visit). The advantage of a fixed number of visits is that the marginal cost of a characteristic can change without changing the conditional visit level. Only the probability of visitation will change.

In aggregating over quantity units, there is, of course, the problem of the period of time over which the aggregation can be carried out. In theory, aggregation can be performed over all quantity units for which the consumer's marginal cost and marginal willingness-to-pay functions remain unchanged. Accordingly, aggregation could cover a weekend, a month, a season, or several seasons, although in practice it is unlikely to go beyond one season.

Identification of Compensated Demand Curve

Even if the assumptions of no repackaging and identical visits are met and the number of quantity units is fixed, there is still the problem of identification and of what is identified. What is observed for typical market goods are their prices and the levels of the characteristics that are embodied in them. The first step in estimating the hedonic model is to regress the commodities' prices on the levels of the characteristics; the marginal, or hedonic, prices for each characteristic are then calculated. These act as the prices in the second-step, simultaneous estimation of the demand and supply relationships. The fact that there are multiple prices in the market for different mixes of characteristics can itself assist identification. Multiple prices occur because of exogenous factors that cause buyers or sellers or both not to be identical. If buyers are identical but sellers are not, hedonic prices sketch out the demand curve for the characteristic. It is worth noting that this is a compensated demand curve. This is so because in order for multiple sellers to coexist, the consumer must be indifferent about buying from any particular seller. So long as the seller's supply curve for quantity units is rising, it is to be expected that there will be multiple sellers in the market.

If sellers are identical but buyers differ, then it is the seller's offer curve— that is, the seller's indifference curve—that is identified. For multiple buyers to exist, the seller must be indifferent as to whom he sells. So long as one buyer does not purchase all of the quantity units available, multiple buyers will exist.

The difficult case is also the more general case. It occurs when neither buyers nor sellers are identical and the hedonic prices do not necessarily trace out either the demand or the supply relationship. The hedonic case is more difficult than the ordinary identification problem because all suppliers and demanders are participating in the market at the same time. This implies a greater endogeneity than normally exists in conventional demand-and-supply models. It also implies that there can be only one price for a given package of characteristics. As buyers and sellers sort themselves out along the hedonic price function, not only will both the price and the package be determined by the exogenous demand and supply shifters, but the latter may well be highly correlated, preventing assessment of the independent effects of either. The solution must either be to impose restrictions on the functional forms of the demand and supply relationships or to introduce new information to provide exogeneity. The use of data from more than one market has been suggested (Bartik and Smith, 1987; Brown and Rosen, 1982; Mendelsohn, 1985).

Even if restrictions on functional form or multiple markets are used, there is still the possibility of the type of simultaneity that is normally found in demand-and-supply models. Whether single-market or multimarket data are used, if the marginal hedonic price for a characteristic depends on the level of the characteristic, the instrumental-variable or a similar approach is required to avoid simultaneous equation bias (Bartik and Smith, 1987).

The recreation case does not present as difficult an identification problem as the general case just described because there can be more than one price for a given package of characteristics. Recreationists from different origins will, in general, face different prices for the same package, which facilitates identification. It does not mean, however, that the identified demand curve will be a compensated one. There is no built-in market mechanism to ensure that a given recreationist will be indifferent regarding sites chosen by himself and by his identical counterparts at other origins.[2]

Although variation in package prices across origins facilitates identification, estimation is not straightforward. First, because the supply curve for a characteristic is determined by the geographic location of sites and not by the market, the shape need not be smooth and upward sloping. Second, the more characteristics whose demand curves need to be identified, the more origins are required.

Recreation Applications

The main applications of the hedonic model to recreation site choices have been those reported by Mendelsohn (1984a,b) for deer hunting, by Brown and Mendelsohn (1984) for fishing, and by Wilman (1983) for deer hunting. The applications differ somewhat in their ways of estimating the cost curves and of treating visits.

Mendelsohn (1984a,b) and Brown and Mendelsohn (1984) estimate cost curves on a per visit basis and do not assume that the visit level is invariant to changes in the cost of a characteristic. Because they use their results only to estimate the value of marginal changes in the characteristic level, the demand curve for visits is not of concern. Wilman (1983), on the other hand, estimates demand curves on a seasonal basis, with the visit level being invariant to changes in characteristic costs. The results are used to estimate the benefit from a shift in the marginal cost curve for a characteristic. If visits had not been invariant, it would have been necessary to estimate a demand curve for visits as well as for the characteristic.

With the exception of Mendelsohn (1984b), the estimation of the cost curves for characteristics in these models has usually been constrained to a functional form that is linear in the characteristic of concern. This facilitates identification of the characteristic demand curves because the hedonic prices can be treated as exogenous in the demand curve estimation. Mendelsohn relaxes this linearity assumption, however, and uses two-stage least squares to estimate a simultaneous system. Nevertheless, his results show that the linearity assumption produces estimates that are reasonably close to the two-stage least-squares results.

Strengths and Weaknesses

The strength of the hedonic model is that it focuses on characteristics of recreation sites. At least some of these characteristics are those that a public resource manager can manipulate, and knowing the benefits that can be provided by such manipulation is very useful information. The approach works best when visits can be treated as exogenous, although this is not as important if one is only interested in the marginal value of the characteristic. The main difficulties of this method are the estimation of the cost curves, the complexities introduced when the estimated hedonic prices cannot be treated as exogenous in the demand curve estimation, and the difficulty of obtaining sufficient exogenous supply variation to identify a large number of characteristics. As Mäler (1974) has pointed out, the identification of a full set of demand equations may be impossible if the number of characteristics is large because geographic variation is limited by the two-dimensional nature of land.

The Travel-Cost Model

The main alternative to the hedonic model is the travel-cost model, which has been in use for some time. (The hedonic model is relatively new.) Harold Hotelling, in a letter quoted by Prewitt (1949), and Marion Clawson (1959) originated the travel-cost approach to measure the value of a recreation site. The method has been extended to deal with substitute sites (Burt and Brewer, 1971; Cicchetti, Fisher, and Smith, 1976; and the Cesario and Knetsch, 1976, gravity model). However, our concern here is with valuing quality changes, and that means that substitute sites must be more explicitly defined. In fact, if cross-sectional comparisons are to be used to value changes in site quality, a visit to one site must be a perfect substitute for a visit to another site, the marginal rate of substitution being determined by the relative amounts of characteristics at the two sites. The recreationist's indifference curves for sites, as well as his or her budget line, will be linear; thus, the recreationist will either choose only one site, or else will be indifferent as to a choice of visits to the two sites. In fact, the recreationist's utility function will be the same as that used in the hedonic model, with the exception that only marginal willingness to pay for characteristics need be the same across visits (the total willingness to pay for a visit will not be the same across visits).

Multiple-Site Travel-Cost Models

The Burt and Brewer model is relatively easy to characterize within this framework. The model normally appears as follows:

$$n_1 = f_1(P_1, P_2, P_3, \ldots P_m)$$
$$n_2 = f_2(P_1, P_2, P_3, \ldots P_m)$$
$$n_3 = f_3(P_1, P_2, P_3, \ldots P_m)$$

$$\cdot$$
$$\cdot \qquad (1)$$
$$\cdot$$

$$n_m = f_m(P_1, P_2, P_3, \ldots P_m)$$

where n_1 through n_m = number of visits made by an individual
recreationist to sites 1 through m
P_1 through P_m = travel-cost prices of visits to
sites 1 through m

Sites 1 through m differ in that they contain different amounts of a set of characteristics. They can be regarded as site types rather than as individual sites. For any given recreationist, P_1 through P_m can be regarded as the minimum costs of visits to sites of type 1 through m, respectively. In a deterministic world with all site types being perfect substitutes, there will be only one site chosen by a given recreationist at a given set of prices. At a different set of prices, a different site would be chosen. However, if the m relationships in equation (1) have the same functional form, only one site is selected, and thus only one of the n_1 through n_m counts of visits to each site is positive at any given set of prices, then the m equations can be replaced by one general equation:

$$n = f(P_1, P_2, P_3, \ldots P_m) \qquad (2)$$

A quality change at a given site can be viewed as changing some of the set of prices at which site types 1 through m can be visited. If site 1 at price P_1 was the recreationist's choice before the change and site 2 at price P_2 was the choice after the change, the benefit to the recreationist is $cdP_2 - abP_1$ in figure 1.

To estimate the demand curves aD_1 and cD_2, it is necessary to vary the price for that site type while holding the site choice constant. Because this will, in general, change the relative prices of the site types, it seems unlikely that the site choice would remain constant. There are two special cases, however. The first occurs when there are only a few site types and changing

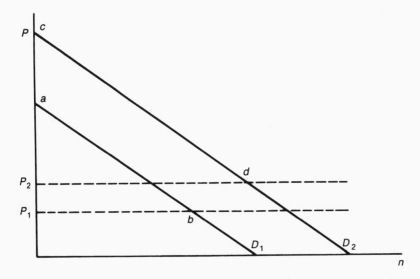

Figure 1. Benefits from a change in the availability of site quality.

the relative prices of the sites will not affect the site choice—at least not over some range. In this case the visits to a site need only be a function of the price of the site and its quality Q.[3] The general demand function can be written as follows:

$$n = (P,Q) \tag{3}$$

where P = the travel cost to the chosen site

Q = the vector of quality characteristics for the chosen site

The second case occurs when the change P_1 also results in changes in P_2 through P_m, leaving the relative prices of site types unchanged. Suppose that sites 1 through m are grouped together in a national forest or national park. Visitors live outside the forest or park and may enter only through a limited number of entry points. Visitors such as a and e in figure 2, coming through different entry points, will face different relative prices. Visitors such as a, b, and c have a different visit price without having different relative site prices. Hence, observations on a, b, and c can be used to obtain a visit demand curve along which site choice is constant. Similarly, e, f, and g can be used to estimate another visit demand curve with a different but constant site choice. These two demand curves are comparable to aD_1 and cD_2. In this view of the world, the site price P_i can be broken down into two parts: $P_i = S + R_i \cdot S$ is the part of the site price that does not depend on site choice, and R_i is the part that does. The general demand curve for the chosen site can now be written as follows:

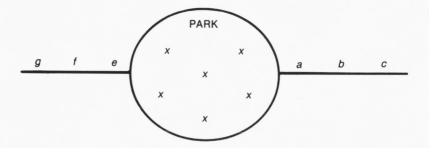

x = sites of different types
a through g = visitors at different locations

Figure 2. Visitor and site distribution for a park.

$$n = f(S, R_1, R_2, R_3, \ldots R_n) \tag{4}$$

where n = the number of visits to the chosen site.

However, if there is no fixed cost that can be changed independently of the relative prices of sites or if there are so many sites that there is no range within which the price of the chosen site can be varied without changing the site choice, then the site choice is likely to adjust to the change in chosen site price so frequently that it will be difficult to identify a demand curve along which site quality is constant.

Travel-Cost Versus Hedonic Model

The hedonic approach works best when the visit level remains fixed as the site choice changes and the marginal willingness to pay for a characteristic and its marginal cost can simply be aggregated over n visits. The travel-cost model works best when the site choice remains fixed as the visit level changes and the marginal cost of a characteristic (or the level of the characteristic) can be used as an exogenous shift variable affecting the demand and supply curves for visits.

Stochastic Elements

The models discussed thus far have been concerned only with deterministic choices. This need not be the case, however, in large part because it is probably unrealistic to assume that a recreationist has perfect information on which to base a decision about whether or not to make a visit and about which site to visit.

The stochastic element in the decision of whether or not to visit a particular site has long been part of applications of the travel-cost approach in which "per capita visits" are a measure of the expected visit level of an individual at a given origin. Ideally it would be best to have data that included not just the expected visit level but also separate measurements of whether or not an individual made any visits and, if so, of the number of visits made. The individuals who choose to make no visits at a given travel-cost price can be viewed as being at a corner solution. Individuals choosing an interior solution at that price will provide the investigator with the conditional visit distribution at that price. An increase in the travel-cost price increases the probability of a corner solution and decreases the probability of an interior solution.

The Limited Dependent Variable Problem

The tendency for more observations to represent corner solution choices as the travel-cost price increases can affect conditional visit distribution. This effect occurs because the switches from interior to corner solutions are not randomly drawn from the conditional visit distribution. Instead, the switches are those observations for which nonpositive conditional visit levels would have been observed, which effectively means that the range of the conditional visit level is limited.[4]

Although in many applications of limited dependent variable models the values for the dependent variables outside the range are meaningful and could have been observed with a different experimental design, nonpositive visit levels will never be observed.[5] Nevertheless, it is true that the limitation in the observable range for the dependent variable, conditional visits, has implications for the estimation of the conditional demand curve of visits and for the consumers' surplus welfare measurement.

The conditional demand curve cannot be estimated by ordinary least squares (OLS) because the limited range of the dependent variable prevents the assumptions of OLS from being realized. As a result, OLS regression coefficients will be biased.

Estimating a limited dependent variable model requires the specification of a form for the conditional demand function and for the discrete choice probability. Hanemann (1984) offers some alternatives. The Tobin (1958) specification, in which the errors are assumed to be normally distributed, is commonly used. A computationally easy two-step estimation procedure has been suggested by Heckman (1976) for the normal error case. However, this two-step approach can also be used with other error specifications (Amemiya, 1985).

Measuring Welfare Gains

Once a limited dependent variable model has been estimated, there is still the question of estimating the welfare gains that accrue to an individual from the availability of the site. From this perspective, some writers are concerned that the specifications of the model be utility theoretic derivations (Bockstael, Hanemann, and Kling, 1987). A more important consideration, however, is that movements between the corner solution and the interior solution, as well as changes in the level of the interior solution, be taken into account. Taking these movements into consideration means that the welfare gains to an individual from being able to visit a recreation site at a lower rather than a higher travel-cost price are conditional on his decision to visit the site and must be adjusted by the probability of a visit to the site. The welfare gains to an individual from having the site available at a given travel-cost price rather than at a price that would reduce his probability of visiting to zero can be measured as the sum of the set of probability-adjusted welfare gains from a series of small price decreases. Because the probability of a visit changes as the price decreases, the probability adjustment will change with the level of the price. Thus, an expected visit locus rather than a conditional demand curve must be used to estimate welfare gains.

The Hedonic Model

The limited dependent variable problem can also arise in the context of the hedonic model. Assume that the visit level is fixed and, consequently, that there is no problem with sample selection bias in the quantity dimension. If, as the travel-cost price of a visit increases and more individuals switch from interior solutions to corner solutions, the individuals who switch are randomly drawn from the quality distribution, then there will be no bias in the quality dimension either. If the individuals who switch are not randomly drawn from the quality distribution, however, the problem of bias again arises.

Site Choice as a Probabilistic Event

The previous discussion leads us to the consideration of site choice itself as a probabilistic event. If site choice were such an event, a recreationist facing a given set of prices and site qualities would have a probability distribution for site choice rather than a certainty of choosing a given site. Changing the set of prices associated with the site qualities would, in general, change the probability distribution. Morey (1985) and Bockstael, Hanemann, and Kling (1987) have applied multinominal logit models to the site choice decision.

Morey assumes a constant visit level (one visit), and his sample includes only those individuals who do make a visit. The probability of visiting a given site is a function of the utility provided by that site relative to all other sites and the cost of visiting that site relative to visiting all other sites. The utility is derived from site characteristics; as a consequence the probability of visiting a site is a function of the levels of a set of characteristics available at that site relative to the levels of the characteristics available at all other sites and the cost of visiting that site relative to the costs of visiting at all other sites.

This model is similar in approach to that of the gravity model mentioned earlier in that the prices and characteristics of all sites are included as predictors. If one accepts the Burt and Brewer (1971) framework, only the least expensive sites of a given type need be included. A restrictive property of the multinominal logit model, however, is the assumption of "independence of irrelevant alternatives," which specifies that the relative probabilities between a pair of alternatives are independent of other available alternatives. In terms of recreation sites the relative probabilities of visiting sites A and B are independent of the availability of other sites. Although this assumption facilitates estimation, it is rather restrictive.

Bockstael, Hanemann, and Kling (1987) used a nested logit model in which the assumption of independence of irrelevant alternatives is no longer required. In the nested logit model the site choice is viewed as a two-step process. Suppose there are two characteristics of campsites: nearness to water (N) and whether or not there are picnic tables (T). The nested logit model allows one to view the choice of a campsite as being a choice of N conditioned on T, and then a choice of T, assuming the best N for every T. (Nested logit models also have their restrictions, which are implicit in the way of nesting.)

In the campsite choice example, suppose there are four alternatives. Alternative A is near water and has a picnic table. Alternative B is far from the water and has a picnic table. Alternative C is near water but has no picnic table. Alternative D is far from the water and has no picnic table. The assumption of independence of irrelevant alternatives says that the relative probabilities of choosing any two of the alternatives does not depend on the availability of the others. The use of a nested model implies that some of the choices are better substitutes for one another than are others. If the way of nesting is that described in the previous paragraph, the campsites are divided into classes according to whether or not they have picnic tables. The relative probabilities of alternative A versus B are independent of the availability of alternatives C and D, and the relative probabilities of the choice of C versus D are independent of the availability of A and B. Yet, relative probabilities across classes are dependent on the availability of other sites within either of the classes. (For example, the relative probabilities of alternative A versus C depend on the availability of both alternatives B and D.)

The restrictions of the nested logit model are similar to the identification restrictions that are required for the hedonic model. For example, T and N would be treated as characteristics in the hedonic model, with the choice of N being conditional on the choice of T. This conditionality implies that the system can be treated as recursive rather than simultaneous, with T being treated as predetermined in the demand-and-supply system for N. Like identification restrictions in the hedonic model, nesting restrictions in the nested logit model must be evaluated in terms of the reasonableness of the structure they impose.

The Participation Decision

Although probabilistic site choice models are intuitively appealing because we observe that a given recreationist does not always make the same site choice, their limitation is that the decision of whether to participate at all, and if so at what level, can only be dealt with indirectly. In general, the continuous part of a discrete/continuous choice is treated as a number of independent discrete choice occasions. The visitation level is determined by adding up the results of these choice occasions.

Ideally, it would be desirable to be able to model the participation decision and the site choice decision together as part of one utility maximization decision. So far, however, this joint modeling has not proven feasible. Bockstael, Hanemann, and Kling (1987) used a two-stage approach. In the first stage, they estimated a nested logit model to explain site choice. From the nested logit model they calculated inclusive values (expected consumer's surplus from a visit), which captured the effect of all of the variables that explain site choice. In stage two, inclusive values are used as explanatory variables in a discrete continuous choice (or limited dependent variable) model.

The two-stage approach is a reasonable one, and it could also be used with hedonic models, which tend to be similarly deficient with respect to the participation decision. However, at least in the case of the hedonic model, the use of the two-stage approach does require that limited dependent variable bias not be present in the quality dimension.

A Simple Repackaging Model

Earlier, in the discussion of the hedonic model, it was mentioned that the no-repackaging assumption does not always fit recreation examples. For recreation activities like hunting and fishing, quality characteristics closely related to the probability of bagging game or catching fish may well be linearly additive across quantity units within some time frame such as a season. For such activities the simple repackaging model may be more appropriate than models that incorporate no-repackaging assumptions. Consider a simple repackaging model for hunting in which the utility function and constraints are given as:

$$U_1(x) + U_2(Q) \tag{5}$$

$$B - x - [h + k(q)]n = 0 \tag{6}$$

$$qn - Q = 0 \tag{7}$$

where B = income or budget constraint
  x = other goods and services
  q = probability of bagging game per visit (assumed to be exogenous at a given site)
  n = number of visits
  h = fixed cost of a visit
  $k(q)$ = variable cost of a visit, which is a function of q
  Q = expectation of bagging game over all visits

The first-order conditions for q and n can be combined to give (8):

$$U_{2Q}/U_{1x} = [h + k(q)]/q = k_q, \qquad Q = qn > 0 \qquad (8)$$

What is implied by this condition is best illustrated diagramatically in figure 3. The variable k_q gives the marginal cost of q per unit of n, and $[h + k(q)]/q$ gives the marginal cost of n per unit of q. Assume that the smallest level n can take if it is to be positive is $n = 1$. Because at $n = 1$ any positive h will result in $k_q < [h + k(q)]/q$, the hunter will first start to increase q to produce more Q. As q increases, k_q increases and $[h + k(q)]/q$ decreases. When q reaches the level at which $k_q = [h + k(q)]/q$, the recreationist is indifferent to the choice between an extra unit of Q obtained by increasing q and an extra unit of Q obtained by increasing n. However, at values past q^*, it will always be less expensive to make more visits. This fact implies that the recreationist will make n_* visits at the q^* site. At the resultant bag expectation level (Q^*), the marginal cost of Q will be $k_{q*} = [h + k(q_*)]/q_*$. It should also be noted that the higher h is, the greater q_* will be and the lower n_* will be. Conversely, at very low levels of h, q_* will be low relative to n_*.

One of the useful implications of this model is that, if multiple visits are made, $[h + k(q_*)]/q_*$ is the marginal cost of Q. This value is simply the marginal cost per unit q of a visit and, because Q and q must be measured in any case, $[h + k(q_*)]/q_*$ is easier to measure than the k_q required in hedonic models.

The Quantity Unit

Particularly in the case of the simple repackaging model, it is clear that if visits are to be used as quantity units, they must be of equal length. Yet it is often observed that visits are not of equal length. One approach to dealing with visits of unequal length is to analyze visits of different lengths separately, as Brown and Mendelsohn (1984) have done. This approach is also easily

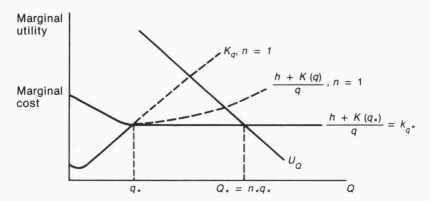

Figure 3. The simple repackaging model.

incorporated into the discrete choice models: visits of different lengths to the same site are different choices. However, this approach does increase the number of choices and hence the difficulty and cost of estimating the model.

An alternative is to specify a production function for recreation services—for example, that specified by Desvousges, Smith, and McGivney (1983). The recreation services provided by a site thus would be a function of the visits to the site and the time spent on site per visit. If a simple repackaging relationship is used, the visits and time spent on site per visit become perfect substitutes, and the measure of the recreation service is the total time spent on site. In such a case the choice between another visit and a longer visit simply becomes a matter of which is the cheaper alternative.

Data from a survey of hunters using the Black Hills National Forest (Wilman, 1983) showed that hunters who lived very close to the forest took many one-day trips, whereas hunters who lived far away took only one long trip. In the intermediate-distance range, there was a switchover region where both one-day visits and longer visits were observed. The survey results suggested the following: (1) at very low travel costs, it was cheaper to make another one-day visit than to increase the length of visit; (2) at high travel-cost levels, it was cheaper to extend the length of visit than to make another visit; and (3) at some intermediate distance the relative costs were the same. These findings are quite consistent with the simple repackaging model if the marginal cost of an extra day added on to the visit increases as visit length increases. If the marginal cost of an extra day increases gradually with visit length, then visit length would increase gradually with the distance from the site. Consider the following utility function:

$$U_1(x) + U_2(d) \tag{9}$$

where x = all other goods and services
d = days at the recreation site for season
$d = \ell \cdot n$
ℓ = length of a visit in days
n = number of visits

The budget constraint is assumed to be of the form $B - x - [h + \gamma(\ell)]n = 0$ where h is the travel cost associated with a visit, and $\gamma(\ell)$ is the opportunity cost of the time associated with a visit. It is assumed that the time used in recreating is fixed across days and that the remainder of the time in a day is spent either traveling or in other activities that yield zero utility. It is also assumed that there is always sufficient time remaining to accommodate travel time. The first-order conditions will include the following:

$$U_{2d}/U_{1x} = \gamma_\ell = [h + \gamma(\ell)]/\ell, \qquad n > 0, \ell > 0 \tag{10}$$

This principle says that the marginal utility of a day equals the per visit marginal cost of an extra day of visit length and that it also equals the per unit length marginal cost of an extra visit. The latter can also be interpreted as the per visit average cost of visit length. It is possible to show the relationship between the marginal and average costs of additional visit length diagramatically (figure 4). Again, it is assumed that the smallest level n can take if it is to be positive is $n = 1$. So long as γ_ℓ is less than $[h + \gamma(\ell)]/\ell$, extra days will be taken by increasing visit length. At ℓ_*, the point at which $\gamma_{\ell_*} = [h + \gamma(\ell_*)]/\ell_*$, the recreationist is indifferent to the choice between an extra day taken as an increase in visit length and an extra day taken as another visit. When γ_ℓ exceeds $[h + \gamma(\ell)]/\ell$, extra days will be taken as additional visits and the marginal cost of each extra day will be equal to the common value of the marginal costs at ℓ_*—that is, $\gamma_{\ell_*} = [h + \gamma(\ell_*)]/\ell_*$. An increase in h will, in general, cause ℓ_* to increase; consequently, at increasing distances, gradually increasing visit lengths and decreasing visit numbers will be observed. If the change in visit length is discrete rather than continuous, it must be expected that $\gamma(\ell)$ will change discretely rather than continuously. For example, suppose that $\gamma_\ell = \gamma_1$ if $\ell \leq \ell_1$ and $\gamma_\ell = \gamma_2$ if $\ell > \ell_1$. The budget constraint then becomes

$$B - x[h + \gamma_1 \ell + (\gamma_2 - \gamma_1)(\ell - \ell_1)]n \qquad (11)$$

Figure 5 is analogous to figure 4 except that the marginal cost functions are discrete rather than continuous. First it is worthwhile to note that, for any positive travel cost, no visit shorter than ℓ_1 will be made because for visits

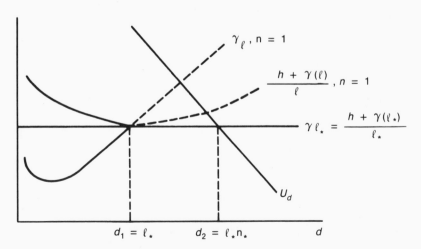

Figure 4. Choice of visit length (continuous case).

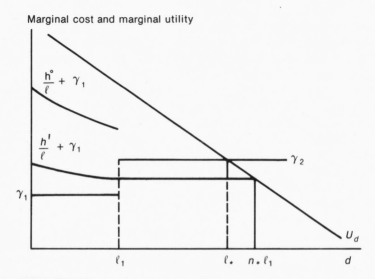

Figure 5. Choice of visit length (discrete case).

up to this length the per visit marginal cost of visit length is less than the per unit length marginal cost of another visit. That is,

$$\gamma_1 < h/\ell + \gamma_1 \text{ when } h > 0 \tag{12}$$

The more interesting question, however, is whether visits longer than ℓ_1 will be made. The answer depends on the size of h. Two cases are shown in figure 5. When h' is the travel cost, then the per visit marginal cost of an extra unit of visit length exceeds the per unit length marginal cost of an extra visit; that is, equation (13) holds and n_* visits of length ℓ_1 will be made:

$$h'/\ell_1 + \gamma_1 < \gamma_2 \tag{13}$$

On the other hand, if $h°$ is the travel cost, the per visit marginal cost of an extra unit of visit length will be less than the per unit length marginal cost of an extra visit. In this case, equation (14) will hold and one visit of ℓ_* length will be made:

$$h°/\ell_1 + \gamma_1 > \gamma_2 \tag{14}$$

At some $h*$ that falls between $h°$ and h' at which $h^*/\ell_1 + \gamma_1 = \gamma_2$, the recreationist will be indifferent to the choice between a longer visit and another visit of length ℓ_1. This point is the switchover point.

Introducing a quality characteristic such as q into the model produces some complications. The choice of q can be expected to be different for recreationists making one long visit and those making many short visits. Because the q choice is made on a per visit basis, visitors making longer visits are likely to choose a larger q than visitors making shorter visits. In the simple repackaging model using $Q = n\ell q$ as the choice variable, the choices of n, q, and ℓ are all related.

Allowing days rather than visits to be the quantity unit in the simple repackaging model and using the discrete switching model of equation (11), the constraints in (6) and (7) become:

$$B - x[h + k(q) + \gamma_1\ell + (\gamma_2 - \gamma_1)(\ell - \ell_1)]n = 0 \qquad (15)$$

$$q1n - Q = 0 \qquad (16)$$

The summary of first-order conditions in equation (8) becomes either (17) or (18):

$$U_{2Q}/U_{1x} = \frac{k_q}{\ell} = \frac{\gamma_2}{q} \qquad Q = n\ell q > 0 \qquad (17)$$

$$U_{2Q}/U_{1x} = \frac{k_q}{\ell_1} = \frac{h + k(q) + \gamma_1\ell_1}{q\,{}_1^\ell} \qquad Q = n\ell q > 0 \qquad (18)$$

Condition (17) suggests that, for visitors making one long visit, the choice of the q and ℓ levels will be such as to equalize the marginal cost of q per day of trip length and the marginal cost of another day per unit of q. In other words, the choice of the q and ℓ levels equalizes the marginal cost of another unit of Q from the two alternative inputs. Similarly, condition (18) suggests that, for visitors taking many ℓ_1-day visits, the choice of q and n levels will be such as to equalize the marginal cost of producing another unit of Q through consuming more q per day and the marginal cost of producing more Q by making another visit.

Whether one visit or many visits of length ℓ_1 are chosen depends on the relative sizes of the right-most terms in (17) and (18), evaluated at the level of q that satisfies the equalities in (17) and (18), respectively.

To measure the marginal cost of q, it is now necessary to measure either k_q/ℓ, or one of γ_2/q or $(h + k(q) + \gamma_1\ell)/q\ell_1$, depending on the choice of ℓ versus n. With γ_1 and γ_2 as part of the latter marginal cost estimates, the estimation of the marginal cost of Q is no longer facilitated by the use of the simple repackaging model. Nevertheless, it is still true that the simple repackaging model with its assumption of linear additivity may be a better description of some recreational choice situations than either the hedonic or the travel-cost models.

Summary and Conclusions

The hedonic, the travel-cost, and the simple repackaging models are all models of recreation site choice that can be useful in valuing quality improvements at recreational sites. The models are best viewed as complements rather than as substitutes, because each has a comparative advantage depending on the particular characteristics of the data set being analyzed. The hedonic model works best if the level of consumption of the quantity unit can be taken as fixed while the site quality choice varies. The simple repackaging model works best when the quality characteristic level is linearly additive across quantity units. Although the approach could apply to other types of quality character-istics, it is most clearly relevant to such characteristics as the probability of bagging game, in which doubling the amount of time spent using a site with a given level of the characteristic can double the effective level of the char-acteristic obtained by the recreationist.[6]

All of the approaches depend on the accurate identification of the quality characteristics and quantity units and the accurate measurement of their levels. Although all of the models postulate that the recreationist's site choice is a function of the quality characteristics available at the various sites, in fact it is often true that our selection of characteristics is determined more by what information happens to be available than by a knowledge of the objective characteristics of sites that appeal to recreationists. If there are to be credible measurements of the benefits that would accrue because the level of a quality characteristic is changed at a site, it is essential that all of the relevant char-acteristics be included and accurately measured.

The identification and measurement of the quantity unit can be important as well, particularly in the case of the simple repackaging model. In this model the level of the characteristic can be increased by increasing the number of quantity units consumed, as well by increasing the level of the quality char-acteristic consumed per quantity unit. The marginal cost of a quantity unit can be important information in identifying the demand curve for the char-acteristic. If the wrong quantity unit is chosen, the marginal cost of the characteristic will be incorrectly measured.

Even if the quality characteristics and quantity units are accurately identified and measured, there are still a number of estimation and identification prob-lems. These problems vary in severity across the approaches. The hedonic approach has one estimation problem that the other approaches manage to avoid: it requires the estimation of the marginal cost curves for the charac-teristics. However, because the cost associated with obtaining a particular bundle of characteristics is not determined by the market but rather by the geographic locations of sites, the marginal cost can be difficult to estimate. If this method is to be used, it is worthwhile to give some consideration to alternative ways of estimating the cost curves. There are two possibilities: (1) to estimate the cost curves using existing data on sites visited by recreationists from a given origin, or (2) to estimate them using the minimum cost of obtaining a given bundle. The first alternative assumes that visitors make their site choice decisions so as to minimize the cost of obtaining the desired bundle. If recreationists at a given origin have different preferences, they will choose different lowest-cost bundles. The variation in this choice identifies the cost function. The second alternative uses data on available sites rather than chosen ones and calculates the minimum cost of obtaining a given bundle. When the minimum costs of obtaining all of the possible bundles are known, cost curves can be calculated.

Although it is not always recognized as such, identification is a problem for all of the models. The hedonic model and the simple repackaging model must identify demand curves for characteristics. The travel-cost model must identify demand curves for visits, and they must be identified at different levels of the various characteristics. If there are a large number of characteristics, this can be a forbidding problem. If the number of characteristics is not too large, however, and simplifications such as recursive models can be used, the problem may be surmountable.

Stochastic variables clearly should be a part of all recreation site choice models. Not only is the decision of whether or not to visit stochastic, but so is the site choice. Ignoring the probabilistic nature of the decision on participation can result in biased parameter estimates for the demand equation and incorrect consumers' surplus estimates. Ignoring the probabilistic nature of site selection is not only intuitively unappealing, but it may also result in biased estimates of the benefits from site quality improvements.

Although it is clear that further work needs to be done, the models that have been discussed in this paper are important because they focus on the question that is particularly vital to the efficient management of public resources. As suggested by Krutilla, Bowes, and Wilman (1983), they provide a methodological base from which the body of knowledge on the valuation of changes in the levels of recreational resource services may be brought into line with information available on the more technical (physical, biological, and engineering) aspects of public land management. John Krutilla has identified the goal and charted a good part of the course. Building up this body of knowledge is a necessary step in making the efficient allocation of public land resources operational.

NOTES

1. Rosen (1974, p. 49) defines the individual consumer's total willingness to pay (or bid) function over n units purchased as $nb(q_1, \ldots, q_m)$, where

$q_1, \ldots, q_m$ are the levels of each of the m characteristics. Because neither $b(z_1, \ldots, z_m)$ nor $b_{z1}(z_1, \ldots, z_m)$ depends on n, both the total willingness to pay and the marginal willingness to pay are the same for each unit of m consumed, and only one type will be purchased no matter how many units are consumed. In goods space, this implies linear indifference curves for goods.

Although it exhibits bid functions that are different from those specified by Rosen, there is an alternative model that will give the conclusion that only one type will be purchased. As discussed later in this paper, Muellbauer (1974) achieves the same conclusion using a simple repackaging model. In this model the bid function is $b(\Sigma z_1 n_i, \ldots, \Sigma z_m n_i)$, where n_i is the number of units of good i that are consumed. This model also exhibits linear indifference curves for goods: either there will be a corner solution in which only one good is consumed, or the distribution of demand among the goods will be indeterminate. The marginal bid function for z_1 is $n_j b_{x1}(z_1 n_j, \ldots, z_n n_j)$, where j is the good selected for consumption and $x_1 = z_1 n_j$. The marginal bid function for z_1 per unit of good j consumed is $b_{x1}(z_1 n_j, \ldots, z_n n_j)$, which is not independent of the number of units of good j consumed.

2. Because the Marshallian rather than the compensated demand curve will be estimated, it will be necessary to test for income effects. If these are significant, an approach such as that suggested by Hausman (1981), which derives the compensated demand curve from the Marshallian demand curve, may be necessary.

3. Essentially, this is the model used by Vaughan and Russell (1982); Desvouges, Smith, and McGivney (1982); and Violette (1985). These varying-parameter models are the most innovative of this type to date because they all allow the estimable parameters of a site demand curve to be a function of quality characteristic.

4. Limited dependent variable models are characterized as being either censored or truncated. In truncated models there is no information on any observation for which the dependent variable is outside the specified range. For censored models there is information on the predictor variables such as the travel-cost price.

5. An example of a case in which a potentially observable variable is not observed is a survival time experiment. Censoring or truncation occurs when a patient is still alive at the last observation time. The survival time is not observed in the experiment but could have been observed had the time of the last observation been later.

6. When such variables as bag and catch probabilities are introduced, there is sometimes concern with regard to how the consumer's choice is affected by bag or catch restrictions. If these restrictions are on a per day basis, they probably have very little effect on the site choice. Instead of the utility that can be derived from a site being a function of the unrestricted catch level

available at the site, it simply becomes a function of bagging the limit at that site. Seasonal restrictions may have a greater influence on site choice, particularly in the hedonic and travel-cost models. In the travel-cost model, a higher-quality site may result in fewer visits because of the bag restriction. In the hedonic model, visits may not remain constant because a higher-quality site means that the bag limit is reached sooner. Only in the simple repackaging model is a seasonal bag limit easily incorporated into the model. If the bag limit is an effective constraint, it simply causes the demand curve for Q to be perfectly inelastic. However, benefits from site quality improvements are easily measured as reductions in the cost of achieving the bag limit.

REFERENCES

Amemiya, Takeshi. 1985. *Advanced Econometrics* (Cambridge, Mass., Harvard University Press).

Bartik, Timothy J., and V. Kerry Smith. 1987. "Urban Amenities and Public Policy," in Edwin S. Mills, ed., *Handbook of Regional and Urban Economics* vol. II (Amsterdam, North-Holland).

Bishop, R. C., T. A. Heberlein, and M. J. Kealy. 1983. "Contingent Valuation of Environmental Assets: Comparisons with a Simulated Market," *Natural Resources Journal* vol. 23, pp. 619–633.

Bockstael, Nancy E., W. Michael Hanemann, and Catherine L. Kling. 1987. "Estimating the Value of Water Quality Improvements in a Recreational Demand Framework," *Water Resources Research* vol. 23 (May) pp. 951–960.

Bowes, Michael D., and John V. Krutilla. In press. *Multiple-Use Management: The Economics of Public Forestlands* (Washington, D.C., Resources for the Future).

Brookshire, D. S., M. A. Thayer, W. D. Schulze, and R. C. d'Arge. 1982. "Valuing Public Goods: A Comparison of Survey and Hedonic Approaches," *American Economic Review* vol. 72, no. 1, pp. 165–177.

Brookshire, D. S., L. S. Eubanks, and A. Randall. 1983. "Estimating Option Prices and Existence Values for Wildlife Resources," *Land Economics* vol. 59, no. 1, pp. 1–15.

Brown, Gardner M., Jr., and Robert Mendelsohn. 1984. "The Hedonic Travel Cost Method," *Review of Economics and Statistics* vol. 66, pp. 127–133.

Brown, James N., and Harvey S. Rosen. 1982. "On the Estimation of Structural Hedonic Price Models," *Econometrica* vol. 50, no. 3, pp. 765–768.

Burt, O. R., and D. Brewer. 1971. "Estimation of Net Social Benefits from Outdoor Recreation," *Econometrica* vol. 39, no. 5, pp. 813–827.

Cesario, F. J., and J. L. Knetsch. 1976. "A Recreation Site Demand and Benefit Estimation Model," *Regional Studies* vol. 10, pp. 97–104.

Cicchetti, Charles J., and V. Kerry Smith. 1973. "Congestion, Quality Deterioration and Optimal Use: Wilderness Recreation in the Spanish Peaks Primitive Area," *Social Science Research* vol. 2 (March) pp. 15–30.

Cicchetti, Charles J., Anthony C. Fisher, and V. Kerry Smith. 1976. "An Econometric Evaluation of a Generalized Consumer Surplus Measure: The Mineral King Controversy," *Econometrica* vol. 44, no. 6, pp. 1259–1276.

Clawson, Marion. 1959. "Methods of Measuring the Demand for and Value of Outdoor Recreation." Paper presented at a meeting of the Taylor-Hibbard Club, University of Wisconsin (Resources for the Future Reprint No. 10).

Court, Louis M. 1941. "Entrepreneurial and Consumer Demand Theories for Commodity Spectra," *Econometrica* vol. 9, no. 1, pp. 135–162.

Davis, Robert K. 1963. "The Value of Outdoor Recreation: An Economic Study of the Maine Woods" (Ph.D. thesis, Harvard University, Cambridge, Mass.).

Desvousges, W., V. K. Smith, and M. P. McGivney. 1982. *A Comparison of Alternative Approaches for Estimating Recreation and Related Benefits for Water Quality Improvements.* Report prepared for the U.S. Environmental Protection Agency by the Research Triangle, Research Triangle Park, N.C.

Fisher, Anthony C., and John V. Krutilla. 1972. "Determination of Optimal Capacity of Resource-Based Recreation Facilities," in John V. Krutilla, ed., *Natural Environments: Studies in Theoretical and Applied Analysis* (Baltimore, Md., Johns Hopkins University Press for Resources for the Future).

Fisher, F. M., and K. Shell. 1971. "Taste and Quality Change in the Pure Theory of the True Cost of Living Index," in Zvi Griliches, ed., *Prices and Quality Change: Studies in New Methods of Measurement* (Cambridge, Mass., Harvard University Press).

Haigh, John A., and John V. Krutilla. 1980. "Clarifying Policy Directives: The Case of National Forest Management," *Policy Analysis* (Fall) pp. 375–384.

Hammack, Judd, and Gardner Mallard Brown, Jr. 1974. *Waterfowl and Wetlands: Toward Bioeconomic Analysis* (Washington, D.C., Resources for the Future).

Hanemann, W. M. 1984. "Discrete/Continuous Models of Consumer Demand," *Econometrica* vol. 52, no. 3, pp. 541–561.

Hausman, J. 1981. "Exact Consumer's Surplus and Deadweight Loss," *American Economic Review* vol. 71, pp. 662–676.

Heckman, J. J. 1976. "The Common Structure of Statistical Models of Truncation, Sample Selection and Limited Dependent Variables and a Simple Estimator for Such Models," *Annals of Economic and Social Measurement* vol. 5, no. 4, pp. 475–492.

Houthakker, H. S. 1952. "Compensated Changes in Quantities and Qualities Consumed," *Review of Economic Studies* vol. 19, no. 3, pp. 155–164.

Krutilla, John V. 1966. "Is Public Intervention in Water Resources Development Conducive to Economic Efficiency?" *Natural Resources Journal* vol. 6, pp. 60–75.

———. 1967. "Conservation Reconsidered," *American Economic Review*, vol. 57, no. 4, pp. 777–786 (Resources for the Future Reprint No. 67).

———, ed. 1972. *Natural Environments: Studies in Theoretical and Applied Analysis* (Baltimore, Md., Johns Hopkins University Press for Resources for the Future).

———, and Anthony C. Fisher. 1975. *The Economics of Natural Environments: Studies in the Valuation of Commodity and Amenity Resources* (Baltimore, Md., Johns Hopkins University Press for Resources for the Future).

———, and John A. Haigh. 1978. "An Integrated Approach to National Forest Management," *Environmental Law* vol. 8, no. 2, pp. 373–415.

———, and Jack L. Knetsch. 1970. "Outdoor Recreation Economics," *Annals of the American Academy of Political and Social Science* vol. 389, pp. 63–70.

———, Michael D. Bowes, and Elizabeth A. Wilman. 1983. "National Forest System Planning and Management: An Analytical View and Suggested Approach," in R. A. Sedjo, ed., *Governmental Interventions, Social Needs, and the Management of U.S. Forests* (Washington, D.C., Resources for the Future).

Mäler, Karl-Göran. 1974. *Environmental Economics: A Theoretical Inquiry* (Baltimore, Md., Johns Hopkins University Press for Resources for the Future).

Mendelsohn, Robert. 1984a. "An Application of the Hedonic Travel Cost Framework for Recreation Modeling to the Valuation of Deer," pp. 89–101 in V. K. Smith, ed., *Advances in Applied Micro-Economics* vol. 3 (Greenwich, Conn., JAI Press).

———. 1984b. "Estimating the Structural Equations of Implicit Markets and Household Production Functions," *Review of Economics and Statistics* vol. 66, pp. 673–677.

————. 1985. "Identifying Structural Equations with Single Market Data," *Review of Economics and Statistics* vol. 67, pp. 525–529.

Mitchell, Robert C., and Richard T. Carson. 1981. "An Experiment in Determining Willingness to Pay for National Water Quality Improvements." Unpublished manuscript (Washington, D.C., Resources for the Future).

Morey, E. R. 1985. "The Logit Model and Exact Expected Consumer's Surplus Measures: Valuing Marine Recreational Fishing." Paper presented at the Association of Environmental and Resource Economists' Workshop on Recreation Demand Modeling, Boulder, Colorado.

Muellbauer, J. 1974. "Household Production Theory, Quality and the Hedonic Technique," *American Economic Review* vol. 64, no. 6, pp. 977–994.

Muth, Richard F. 1966. "Household Production and Consumer Demand Functions," *Econometrica* vol. 34, pp. 699–708.

Prewitt, R. E. 1949. "Economic Study of the Monetary Evaluation of Recreation in National Parks," unpublished paper. U.S. Department of the Interior, Washington, D.C.

Rosen, Sherwin. 1974. "Hedonic Prices, Implicit Markets and Product Differentiation in Pure Competition," *Journal of Political Economy* vol. 82, pp. 34–50.

Tinbergen, J. 1956. "On the Theory of Income Distribution," *Weltwirtschaftliches Archiv* vol. 77, pp. 155–173.

Tobin, J. 1958. "Estimation of Relationships for Limited Dependent Variables," *Econometrica* vol. 26, pp. 24–36.

Vaughan, W. J., and C. S. Russell. 1982. "Valuing A Fishing Day: An Application of a Systematic Varying Parameter Model," *Land Economics* vol. 58, no. 4, pp. 450–463.

Violette, Daniel M. 1985. "A Model to Estimate the Economic Impacts on Recreational Fishing in the Adirondacks from Current Levels of Acidification." Paper presented at the Association of Environmental and Resource Economists' Workshop on Recreation Demand Modeling, Boulder, Colorado.

Wilman, Elizabeth A., with the assistance of Paul Sherman. 1983. "Valuation of Public Forest and Rangeland Resources." Discussion Paper D-109 (Washington, D.C., Resources for the Future).

PUBLIC INTERVENTION REVISITED: IS VENERABILITY VULNERABLE?

Charles W. Howe

In 1966 John Krutilla published an article entitled "Is Public Intervention in Water Resources Development Conducive to Economic Efficiency?" This article, which sharply pointed out many abuses of standard economic analysis that had crept into the practices of the federal agencies responsible for water development, dealt with three key issues: (1) the presence of "government failure" as well as market failure in the economy; (2) the importance of reimbursement policy in stimulating socially appropriate decisions by the economic and political actors associated with federal water programs; and (3) the inability of federal agencies to adapt to a changing economic, political, and technological environment. This last topic was introduced under the heading "The Vulnerability of Venerability," reflecting the view current at the time that the early 1960s seemed to be a period of rapid change in U.S. water policy and practice. As our review of the intervening period will show, the venerable practices of the water agencies appear not to be as vulnerable as one might have judged in the 1960s.

Krutilla's article ends with a plea for institutional innovation that emphasizes the centrality of reimbursement policy in motivating socially responsible behavior. As he states in its introduction:

Attention is directed towards the institutional mechanism through which public intervention seeks to influence the decision regarding the extent and character of governmental resource development programs, and whether or not these institutions are compatible at present with the efficiency objectives that justify public intervention. (Krutilla, 1966, p. 60)

Krutilla also could well have questioned the compatibility of these institutions with the equity objectives that historically have motivated public programs, although equity was not the topic of the paper.

Krutilla's professional work has been devoted to the improvement of *public policy and programs*, first through his work with the Tennessee Valley Authority (TVA); then through major contributions to the welfare-theory foundations of applied benefit–cost analysis in "Welfare Aspects of Benefit–Cost Analysis" (Krutilla, 1961), *Multiple Purpose River Development: Studies in Applied Economic Analysis* (Krutilla and Eckstein, 1958), and in his work with Hufschmidt and coauthors (1961) on the Bureau of the Budget panel of consultants in *Standards and Criteria for Formulating and Evaluating Federal Water Resources Development*. Somewhat later, his work focused on institutional design and performance (see, for example, *The Columbia River Treaty: The Economics of an International River Basin Development* [Krutilla, 1967]).

In his 1966 article Krutilla foresaw much of the more recent controversy over government agency and private market failure, as evidenced by his statement that "the justification for public intervention must be judged by the relative degree of inefficiency under private as compared with public development" (p. 64). Research since that time has confirmed the inefficiencies found in public water programs, although the privatization literature (for example, Anderson [1983]; Deacon and Johnson [1985]) has certainly failed to make its case that the private market will always be more efficient when dealing with water and other natural resources. Nevertheless, Young and Martin (1967) showed convincingly that there was no economic justification for the Central Arizona Project. Haveman (1972) found numerous projects of the U.S. Army Corps of Engineers that had not fulfilled their investment expectations. Howe and coauthors (1969) found large economic inefficiencies in the operation and expansion of the U.S. inland waterway navigation system. Kelso, Martin, and Mack (1973) studied the efficiency of water use in rapidly growing Arizona, showing the importance of reallocations of water from irrigated agriculture to evolving uses and, implicitly, throwing serious doubt on the need for new water supplies. Howe and Easter (1971), studying large-scale interbasin transfers of water in the American West, demonstrated that a multifaceted program designed to better conserve, reallocate, and manage water was more efficient than additional large water storage and transfer projects.

In a more popular mode, Berkman and Viscusi (1973) presented convincing evidence of the economic inefficiency and environmental damage resulting from the Bureau of Reclamation's program of water development. Also, a group of national conservation organizations headed by the American Rivers Conservation Council (1973, 1977) published two reports entitled *Disasters in Water Development* (volumes 1 and 2). These reports listed prospective Corps of Engineers and Bureau of Reclamation projects that, in addition to being economically unjustified, appeared to endanger irreplaceable natural and cultural resources.

Finally, the National Water Commission (1973) found many inefficiencies in what was then current water planning and development practice and recommended extensive changes in national water policy. Given this general climate of criticism, there should have been little surprise when in February 1977 the new, conservation-oriented Carter administration published a "hit list" of water projects that the administration wanted stopped and placed high priority on major reforms in national water policy.

It thus appears that there was ample justification for Krutilla's concern about the appropriateness of the institutions charged with federal water development and their practices. The remainder of this paper considers the events in the field since Krutilla's 1966 article to see the extent to which similar public agency shortcomings have continued. The paper focuses especially on reimbursement policy, the area Krutilla identified as crucial to the long-run efficiency of the federal water program.

Some Useful Concepts

Before tracing recent developments in federal reimbursement policy and practice, it is useful to review the meaning and relationships among a set of relevant concepts: (1) economic efficiency, (2) project output pricing, (3) revenue generation, (4) cost allocation, and (5) reimbursement policy (or cost sharing).

The concept of economic efficiency underlies nearly all of microeconomics (that is, except purely descriptive studies), and it takes two major operational forms: Pareto efficiency and benefit–cost analysis (see Dorfman and Dorfman [1972, Introduction]). Pareto efficiency is the purest concept, but, unfortunately, it is also the least applicable; its premise is that an economy is efficient if there is no way to reallocate resources to benefit one party without leaving any other party worse off. This measure involves fewer value judgments than benefit–cost analysis, but it is seldom applicable to real-world decisions, which typically benefit some people while damaging the interests of others.

Benefit-cost analysis moves beyond the Pareto concept by adding monetizable benefits and costs across all affected persons to determine whether or not the present value of aggregate benefits from a proposed project or program exceeds the present value of project costs. In principle, benefit–cost analysis can be applied to any project. Implicit in this process, however, is the assumption that a dollar of measurable benefit or cost has the same social significance regardless of who receives the benefits or who bears the costs. Although a number of schemes have been proposed for weighting benefits and costs according to the affected group (see, for example, Squire and van der Tak [1975]), there is little explicit agreement among policymakers on what those weights should be—except, perhaps, in the case of the "benefit of the doubt" that is frequently given projects benefiting disadvantaged groups.

It is important to note at this point that the techniques for measuring benefits and costs have been greatly improved in the past decade (for example, Cummings, Brookshire, and Schulze [1986]); as a result, the quality of measurements taken in the field appear to be improving significantly—with some notable exceptions. Thus, the criticisms of this paper are not aimed primarily at benefit–cost estimation but rather at the sleight-of-hand that often surrounds the project cost accounting and related cost repayment requirements.

Most water projects are multiple purpose projects; that is, they have multiple outputs, supplying municipal water and hydroelectric power and also offering flood control, navigation, and recreation and other aesthetic services. In a multiple purpose project (for example, the reservoir itself), some features contribute simultaneously to the production of all outputs; other features are clearly associated with a particular output (for example, delivery canals with irrigation or urban water supply, turbines and penstocks with power generation, and locks with navigation). The costs of the former project features (those contributing to all outputs) are called *joint costs*; the costs of the latter are called *separable costs*. For any particular project design, there is no logical way to sort out what portion of the joint costs is attributable to any particular output. (Of course, if the project design is varied so as to produce more of one specific output, the incremental joint costs can then be determined.)

Interestingly, the process of designing a project and its operating rules so as to maximize its net economic benefits does not require the allocation of the joint costs to the several outputs. It requires only a knowledge of the incremental joint costs and the separable costs. Optimum design subsumes the economically efficient operation of the project once it is built; thus, users must be induced to use the project outputs until the short-run marginal benefit to them just equals the short-run costs to the system of providing that added output. The efficient operation assumed by an optimum design also means providing a guarantee that those who value the output most highly will actually get it. Usually it is implicitly assumed that, to induce this behavior among users, project management will set the price of each output at each point in time so that the price will be equal to the short-run marginal cost of that output. Naturally, such a manager would have to be acting in the broad public interest to maximize total net benefits rather than trying to maximize profits. Thus, in theory, marginal cost pricing plays an important role in the efficient design and operation of public projects.

Unfortunately, the pricing of a given product is actually expected to achieve at least three major objectives: (1) economic efficiency in design and operation, (2) equity toward those using the outputs, and (3) the generation of revenues for the project authority, which usually must provide at least a part of its own operating funds. These objectives are frequently in conflict. For example, it is almost certain that marginal cost pricing (as described in the preceding paragraph) will not produce sufficient revenue to cover the full costs of an expanding water system, especially in those instances in which a water system is required to be totally self-funding. Thus, real-world pricing schemes are often a compromise among these various objectives.

The history of federal water project administration has reflected the attempt to fulfill these several objectives: full reimbursement of costs for hydroelectric power and municipal water supply, subsidized water for the family farm, and no charge for public goods like fish and wildlife enhancement or for "hard-to-collect" services like flood control. The federal reclamation program started with a limited reclamation fund that was to be replenished through user repayment (originally over ten years). Consequently, to serve these frequently conflicting objectives, certain reimbursement or cost-sharing policies evolved over time in an ad hoc manner.

Reimbursement policy dictates the fraction of the costs attributable to a particular output that must be repaid to the government agency involved. But reimbursement involves more than just the direct pricing of outputs: it can also involve fixed fees paid by output users and property taxes paid not only by direct beneficiaries but by all residents within a conservancy district that may have been formed to contract with the federal agency. Examples of existing reimbursement policies would be the Bureau of Reclamation policy of (presumably) requiring irrigation users to repay all operating, maintenance, and replacement (OM&R) costs of the system, as well as repaying capital costs interest-free over forty years; charging full cost for urban water supply and hydroelectric outputs; and not requiring any reimbursements for flood storage in Reclamation Bureau reservoirs. (For an excellent history of reimbursement policy since 1902, see Burness and coauthors [1980].)

Because reimbursement policy is based on the total costs attributed to a particular output, there must be some way of dividing joint costs among the outputs (separable costs are more easily related to the outputs). This apportionment process, which is inherently arbitrary, is called *cost allocation*. There are several different kinds of cost allocation procedures used in practice; the most popular is the "separable-costs-remaining benefits" (SCRB) method.

As is described in the remainder of this paper, reimbursement policy has been quite flexible over time, and nearly always flexible downward, that is, in the direction of smaller and smaller reimbursements to the federal government. (Again, see Burness and coauthors [1980] for an account of the colorful evolution of these policies of the Bureau of Reclamation.)

The Recent Evolution of Reimbursement Policy in the United States

The Bureau of Reclamation is fond of telling Congress and the American public that the irrigation, municipal–industrial water supply, and power components of its projects reimburse their costs to the federal treasury. This may be true in a bookkeeping sense in which undiscounted monetary payments using monies of diminishing purchasing power eventually add up to the costs allocated to these functions. In terms of current values, however, such a claim is just not valid. In fact, the history of the Bureau of Reclamation's reimbursement policy is a continuing search for new ploys to reduce real reimbursement.

The study by North and Neely (1977) is one of the best-known research efforts relating to effective reimbursement. Table 1 presents a condensation of their results. Navigation and irrigation have the lowest real repayment (that is, in constant dollars appropriately discounted), but none of the purposes noted in the table even comes close to repaying its costs in real terms. The U.S. average for capital costs was 24 percent in 1974; it was 58 percent for operations, maintenance, and replacement.

A more recent study by Franklin and Hageman (1984) of nineteen Bureau of Reclamation projects in the Upper Colorado and Missouri River Basins compares the allocation of the project costs that appeared in the definite plan statements to actual repayment (the latter has been projected to the end of the project's life on the basis of actual payments to date). In this study, the researchers made no adjustments to constant dollars, nor did they use discounting; yet they projected repayment of capital costs to be only 3.8 percent. Regarding the reimbursement of operation, maintenance, and replacement costs, Franklin and Hageman found that only 9 percent of the projected costs would be paid back, in part because large acreages included in the various projects were never developed and in part because of a massive shift in costs from the reimbursable category to the nonreimbursable. The feasibility reports for these projects allocated 69 percent of OM&R to irrigation, 23 percent to

Table 1. Nonfederal Effective Cost-Sharing Rates for All Federal Water Programs and Projects in the United States, 1974
(percent)

Purpose	Capital costs	OM&R[a]	Composite
Urban flood damage reduction	17	40	20
Agricultural water supply	11	92	19
Municipal–industrial water supply	62	78	64
Water quality management	31	98	60
Recreation	16	34	19
Navigation	7	8	7
Hydropower	64	56	64
U.S. weighted average	24	58	30

Source: Adapted, by permission of the American Water Resources Association, from Ronald M. North and Walter P. Neely, "A Model for Achieving Consistency for Cost-Sharing in Water Resource Programs," *Water Resources Bulletin* vol. 13, no. 5 (October 1977) pp. 995–1005.

[a]Operation, maintenance, and replacement costs.

other reimbursable purposes, and 8 percent to nonreimbursable purposes. The projection of actual OM&R repayment among purposes, however, was 7 percent to irrigation, 29 percent to other reimbursable purposes, and 64 percent to nonreimbursable purposes.

In 1981 the General Accounting Office (GAO) reviewed six federal irrigation projects to determine the extent to which projects actually repaid their costs to the federal treasury. The results are given in table 2. A comparison of the first two columns in the table clearly shows the effects of appropriate discounting: historical costs are repaid far in the future, generally with no interest on the unpaid balances.

The GAO study also compared the actual acre-foot payments by irrigators in each of the study projects with estimates of the full cost of providing the water. These results are given in table 3. The figures in table 3 overstate the actual payments per acre-foot, however, because much of the repayment comes from property taxes imposed on residents of the irrigation or conservancy districts. These districts frequently include towns with large tax bases.

In 1983 the Congressional Budget Office carried out a special study of nonfederal cost sharing in federal projects under twenty-five federal agencies, including the Corps of Engineers, the Bureau of Reclamation, and the Soil Conservation Service. The study compared the nominal (or presumably mandated) reimbursement rate with the effective (present value) reimbursement rate for composites of projects carried out by the agencies named above (see table 4). Some of the cases show average effective rates that are far below the

Table 2. Extent of Federal Subsidy to the Capital Costs of Six Irrigation Projects

Project area	Construction costs eventually repaid ($ millions)	Present value ($ millions)[a]	Subsidy ratio (percent)[b]
Auburn-Folsom	724.5	39.8	94.5
Dallas Creek	16.2	1.3	92.2
Fryingpan-Arkansas	88.0	4.4	95.0
North Loup	131.6	2.9	97.8
Oroville-Tonasket	38.9	3.0	92.3
Pollock-Herreid	34.6	1.3	96.4

Source: U.S. General Accounting Office, *Federal Charges for Irrigation Projects Reviewed Do Not Cover Costs*, PAD-81-07 (Washington, D.C., 1981, table 6, p. 37).

[a]The present value of eventual repayments as of the beginning of operations, using a 7.5 percent discount rate.

[b]Subsidy ratio = 1.0 minus the ratio of the present value to nominal construction costs × 100.

Table 3. Comparison of Prices Actually Paid per Acre-Foot of Water with the GAO-Calculated Full Cost for Six Irrigation Projects
(dollars per acre-foot)

Project area	Price paid[a]	Estimated full cost[b]
Auburn-Folsom	19.58	86.54
Dallas Creek	4.24	103.04
Fryingpan-Arkansas	10.08	54.24
Oroville-Tonasket	19.71	129.11
North Loup	12.07	117.93
Pollock-Herreid	11.61	130.50

Source: U.S. General Accounting Office, *Federal Charges for Irrigation Projects Reviewed Do Not Cover Costs.* PAD-81-07 (Washington, D.C., 1981).
[a]O&M costs plus "ability to pay" toward capital cost.
[b]Full capital and O&M cost, calculated with a 7.5 percent discount rate.

minimum mandated rates. In fact, all of the tables presented thus far show that there are many discrepancies between the nominal rates that are often quoted to the public and the effective rates of repayment, that is, repayment expressed in dollars of constant purchasing power that have been properly discounted. In addition, none of these calculations takes into account the effects of inflation, an important factor in further reducing real reimbursement of costs to the federal government.

Naturally, the low rate of real reimbursement tends to make such projects much more attractive to the project area and state because most of the project's benefits accrue to the area in which it is sited. This effect can be seen in the *ex post* cost data from the Colorado–Big Thompson project (C–BT), the water from which is administered by the Northern Colorado Water Conservancy District (NCWCD). The last two lines of table 5 show the tremendous differences between national and project regional costs, calculated in appropriate present-value terms.

Table 5 also brings out the general lack of clarity in the cost concepts frequently presented to the public—for example, the concept of dollars of differing purchasing power simply summed over time versus present values expressed in dollars of constant purchasing power. The table also identifies the escalation of costs that typically takes place over the planning and construction periods of water projects. The initial construction cost estimate was $44 million, but this amount escalated to a reported final construction cost of $163 million. The increased final cost is largely a meaningless figure, however, because it includes dollars of construction expenditure from 1938 price levels to 1953 price levels. In addition, the $163-million figure is not discounted or

Table 4. Nominal and Effective Nonfederal Reimbursement Rates by Project Purpose for Three Federal Agencies
(percent)

Purpose	Corps of Engineers		Bureau of Reclamation		Soil Conservation Service	
	Nominal	Effective	Nominal	Effective	Nominal	Effective
Urban flood-damage reduction	0–50	17	—	—	—	—
Rural flood-damage reduction	0–50	7	—	10	O&M	27
Irrigation	50	19	100	18	50 + O&M	54
Municipal–industrial supply	100	54	100	71	50 + O&M	100
Hydropower	100	61	100	65	—	—
Water quality	0	3	25 + O&M	82	—	—
Fish and wildlife	0–50	11	25 + O&M	13	50 + O&M	57
General recreation	0–50	17	50 + O&M	18	50 + O&M	63
Navigation	0–50	20	—	37	—	—

Note: Dash = not applicable.
Source: Congressional Budget Office, Current Cost-Sharing and Financing Policies for Federal and State Water Resources Development. Special study (Washington, D.C., July 1983, tables 3, 4, 5, 6).

Table 5. Summary Cost Data for the Colorado–Big Thompson (C–BT) Project and the Northern Colorado Water Conservancy District (NCWCD)
(millions of dollars)

Item	Cost
Original construction cost estimate, 1937[a]	44.0
Revised Bureau of Reclamation estimate, 1946	128.1
Revised Bureau of Reclamation estimate, 1952[b]	162.6
Reported final project cost	163.0
Total C–BT project costs from 1937 through project completion in 1953, in 1960 dollars[c]	443.3
Total national C–BT and NCWCD construction and operation costs through 1980, in 1960 dollars[c]	550.7
Project region total, C–BT and NCWCD costs through 1980, in 1960 dollars[d]	107.9

Source: Charles W. Howe, Dennis R. Schurmeier, and William D. Shaw, Jr., "Innovations in Water Management: An Ex-Post Analysis of the Colorado–Big Thompson Project and the Northern Colorado Water Conservancy District," Department of Economics, University of Colorado. June 1982. Photocopy.

[a]U.S. Congress, Senate, 75th Cong., 1st sess., "Colorado–Big Thompson Project," S. Doc. 80 (Washington, D.C., Government Printing Office, 1937).

[b]Bureau of Reclamation, "Colorado–Big Thompson Project; Addendum to Definite Plan Report" (Washington, D.C., 1952).

[c]Costs indexed to 1960 dollars and discounted or compounded to 1960.

[d]Includes actual payments made by parties in the project region, including payments for electric power.

compounded to a common date. If all the construction dollars for the project are adjusted to 1960 price levels (using the *Engineering News Record* construction cost index) and compounded to 1960, the resultant figure is $443.3 million. If all costs are accounted for through 1980, including the external costs imposed on the Lower Colorado River Basin, then the result, expressed in 1960 dollars and discounted to 1960, is $550.7 million. Yet the costs that must be paid by the project region, similarly expressed in 1960 dollars and discounted to 1960, total $107.9 million.

What financial arrangements permit the project region to escape such a large share of the project costs? In this particular case the NCWCD initially had negotiated an upper bound of $25 million as its share of the capital costs of the C–BT project (approximately half the costs were allocated to power generation). In 1937, in the face of an estimated total project cost of $44 million and with the country in the midst of a severe depression during which price levels had fallen, this was probably a reasonable arrangement. Because of postwar inflation over the forty-year repayment period, however—a period that began only when the project was completely in operation in 1957—the real value of the $25-million repayment has been greatly reduced.

There are a number of additional characteristics of the Bureau of Reclamation's reimbursement policy that greatly reduce the real value of reimbursements:

• the creation of so-called reimbursable project purposes (for example, hydropower, irrigation, and municipal–industrial water supplies) and nonreimbursable purposes (such as flood control, recreation, and fish and wildlife) that can bias the allocation of costs toward the nonreimbursable purposes (see Franklin and Hageman [1984]);

• basing irrigator payments on the ability to pay rather than on the costs allocated to irrigation;

• a system of repayment for irrigation costs that includes a ten-year grace period followed by a forty-year repayment period, all without interest;

• the use of the "basin account" concept, a bookkeeping fund of past and anticipated future electric power receipts that presumably "pays" those irrigation costs that are beyond the irrigators' ability to pay;

• allocating project costs to prospective irrigation developments that have not occurred and that seem unlikely ever to occur under the Pick–Sloan development of the Missouri River;

• proceeding with the construction (and even completion) of projects without having contract commitments for the sale of water; and

• the lack of contractual allowances for contract renegotiation under changed cost and price level circumstances.

These financial features have evolved over time, mostly by administrative direction within the bureau and without open congressional consideration. Their net effect has been to reduce greatly the incidence of costs on water users. These features are also likely to be applicable (under a "grandfather"-type policy) to formerly authorized projects unless the projects are renegotiated under more recent financing procedures.

In 1978 President Carter recommended new cost-sharing arrangements for federal water projects: (1) states would be required to contribute "up-front" cash for 10 percent of the construction costs (5 percent if no vendible outputs were produced); (2) revenues from vendible outputs (for example, water supply, irrigation, and power) would be divided between federal and state governments in proportion to their cost shares in construction; and (3) all flood control programs would require a 20 percent nonfederal contribution (Congressional Budget Office, 1983, pp. 7–8). These ideas were not enacted into law.

Under President Reagan the Council on Environmental Quality made recommendations on cost sharing with an eye to promoting new projects (table 6). These recommendations were not released separately but were included in

Table 6. Proposed Post-1983 Cost Sharing for New Federal Water Projects
(percent)

Project purpose	Up-front nonfederal share of construction costs
Hydropower	100
Municipal–industrial water	100
Flood control	35
Recreation[a]	50
Commercial navigation[b]	75
Irrigation	35
Beach erosion	50

Source: Congressional Budget Office, Current Cost-Sharing and Financing Policies for Federal and State Water Resources Development, Special study (Washington, D.C., July 1983, table 2, p. 9).

[a]Can be repaid over time.

[b]The cost sharing for commercial navigation includes 25 percent repayment over time with 50 percent of the costs up front.

a proposed 1983 bill (S. 1031) relating to the Corps of Engineers. The bill that was eventually passed, the Water Resources Development Act of 1986 (P.L. 99-662 [H.R. 6]), has now formally set reimbursement policy for the Corps of Engineers. For harbor work the nonfederal capital cost share runs from 10 percent to 50 percent, depending on project depth, plus an additional 10 percent to be paid over a period not exceeding thirty years. Operating, maintenance, and replacement costs are to be 100 percent federal. For inland waterway transportation, the federal capital cost share is to be 50 percent, payable only from the Inland Waterways Trust Fund that accumulates from fuel taxes imposed on boats using the inland waterways. (The fuel tax is currently 10 cents per gallon, but it will rise to 20 cents per gallon by the year 2000.) This reimbursement provision will place a stringent constraint on federal waterway expenditures, and it is hoped that it will cause waterway interests to set priorities collectively and carefully. The federal share of OM&R costs is to be 100 percent.

For structural flood control measures the nonfederal share is to consist of a cash payment during construction of 5 percent of the separable costs plus the provision of all lands, rights-of-way, and resettlement and a share of the joint costs assigned to flood control. The total nonfederal share is limited to a minimum of 25 percent of capital costs and a maximum of 50 percent. For nonstructural measures the nonfederal share is to be 25 percent, which includes the costs of contributed lands.

Finally, for hydroelectric power and municipal–industrial water supplies, the nonfederal share of both capital and OM&R costs is to be 100 percent; for agricultural water supplies and storm damage protection, 35 percent; and for recreation, approximately 50 percent, depending on the extent of navigational recreation.

It will be interesting to see whether or not these repayments occur in fact, although the law does require a contract with nonfederal interests guaranteeing payment before projects are begun. No mention is made in the law of adjusting for inflation, but postconstruction capital balances carry interest at long-term Treasury rates.

The Bureau of Reclamation currently has no formal cost-sharing guidelines. In specific cases like the Animas-LaPlata project, the Secretary of the Interior has been authorized to negotiate and accept cost-sharing arrangements. The Bureau of Reclamation is currently negotiating with the states of Colorado and New Mexico and the Ute Indian tribes over updated cost sharing for the Animas-LaPlata project in southwestern Colorado. The costs of the project escalated from $336.4 million in 1980 to $614.5 million in September 1984; yet in a period of tumbling agricultural prices, the benefits also escalated and somehow have kept escalating, keeping the bureau's benefit–cost ratio at 1.4. At first, the bureau proposed a $200-million local cost share, but the states resisted that amount. At the moment no final agreement has been reached. Regardless of these arrangements, however, all Bureau of Reclamation projects must still pass the benefit–cost test and produce acceptable environmental impact statements.

Yet, new evasions of cost responsibility continue to occur. For the first time in reclamation history, the federal Animas-LaPlata project includes water delivery in pressurized pipes to the farm headgate; thus, the distribution costs also benefit from federal subsidies, in part because no local district could afford to pay the full costs of Animas-LaPlata water distribution. In addition, the municipal–industrial water supply has been broken into "blocks" so that the town of Durango will start repayment on only those blocks that it has begun to use, rather than on the total municipal–industrial water supply cost. A large part of the total project benefits are listed as municipal–industrial benefits, even though the project is clearly far ahead of Durango's water needs.

In California's Central Valley project, the repayment contracts of all irrigation units have been redated each time a new irrigation area has been brought into operation. This policy has had the effect of extending the repayment period of the earlier areas to seventy years—at no interest (U.S. General Accounting Office, 1985). Since the 1940s municipal–industrial revenues have been insufficient to cover the project's annual interest repayment; since 1982 irrigation revenues have been insufficient to cover annual operating costs.

Bureau of Reclamation policy presumably has been to require a firm repayment contract prior to the construction of municipal–industrial supply facilities. Yet in the Bonneville unit of the Central Utah project in the late 1970s, the bureau realized that the contract with the Central Utah Water Conservancy District would not recover all of the estimated Bonneville unit costs allocated to this purpose. To continue construction "legally," the bureau "deferred" repayment of the costs associated with a portion of the municipal–industrial water supply, invoking the Water Supply Act of 1958. The bureau also reassigned property tax revenue intended for irrigation reimbursement to municipal–industrial repayment. In the opinion of GAO officials, these actions were legally improper (U.S. General Accounting Office, 1985, p. 4).

In 1984 the bureau modified its cost allocation procedure for the Bonneville unit, which resulted in a large increase of costs allocated to irrigation and an equal decrease in the allocation to power. Because the repayment capacity of irrigators was not recalculated, the action left the added irrigation costs to be "repaid" from the Colorado basin account (U.S. General Accounting Office, 1985, pp. 3–6).

In the area of hydroelectric power, Congress has also shown its lack of concern with repayment economic efficiency, and environmental quality in the matter of pricing Hoover Dam power. Over the first forty-five years of the dam's operations, the average price of its energy was less than 1.5 mills (0.15 cents) per kilowatt-hour. In 1983 the average price was still less than 5 mills (0.5 cents) per kilowatt-hour. (By comparison, market prices for firm power in the Southwest ranged from 2 cents to 7 cents.) According to testimony by Thomas J. Graff on Hoover Dam power marketing (March 6, 1984, before the House Subcommittee on Water and Power of the Committee on Interior and Insular Affairs), even the fuel-cost-averted price paid for co-generated power by private utilities averaged around 4.4 cents per kilowatt-hour.

The prices stated in the original 1935 Hoover Dam contracts were not renegotiable, and they were mandated for fifty years, to expire in 1987. In 1985 Congress had the opportunity to redress the inefficiencies caused by such drastic underpricing of Hoover power to the select few by pricing Hoover power closer to market values. By this action, Congress could have prevented the construction of several new coal-fired, thermal-electric plants, leading the way to much more reasonable power consumption patterns in an era of environmental and energy concern. Unhappily, Congress extended the original contract for another thirty years.

Western Water Policy and the National Interest: Is Venerability Vulnerable?

It seems quite clear from the record detailed in the preceding sections that federal water programs have found ways of reducing the real reimbursement of costs to the federal government. The continuation of these policies is a joint product of agency motivation, beneficiary-group influence, and the concurrence and encouragement of Congress and state governments. This picture is quite consistent with the statement by Shabman (1985, p. 6) that "debates over government policy are unabashedly normative discussions about preferences people ought to hold and the fair distribution of the nation's wealth, issues that lie outside the mainstream of economic analysis."

These recent maneuvers show that, indeed, new facets are being added to the old policies, that is, that venerability is not yet proving vulnerable. In a period of sharply diminishing federal resources (that is, for all purposes but defense), the socially responsible allocation of federal resources should be receiving greater emphasis and not less. Is there hope for reform, and if so, what shape might that reform take?

Let us try to characterize ideal decision making—ideal in terms of the openness of the decision process to the use of good technical information on the economic, environmental, and social effects of proposed projects or programs. We invoke this notion of the ideal because as social scientists we are trying to improve the quality and accountability of public-sector decisions. Yet our only hope of being helpful is if the political actors in the process want good technical input. What would such a decision setting be like?

First, the decision process would use an explicit multiple-objective framework that recognized the relevance of various objectives to social welfare: quantifiable economic benefits and costs, the distribution of these benefits and costs among social groups and locations, environmental effects that cannot be adequately monetized and included among the economic benefits and costs, and similar social impacts. The recognition of more than one objective opens up the possibility of accepting projects that may not seem favorable in terms of some objectives but that fulfill others quite well (for example, political decision makers might accept a project with a negative present value of economic net benefits, provided income is redistributed in a socially desirable direction or the environmental preservation effects of the project are outstanding). The hope is that this type of decision setting would eliminate much of the pressure to distort the analysis of particular objectives—as benefit–cost analyses are currently distorted for water projects. Under multiple-objective decision making, if a project accomplishes explicit social or environmental goals at a cost of net economic losses, then it may be defended in terms of those goals with full knowledge of the extent of its economic losses.

Second, the results of the analyses of the various objectives associated with a project should be freely available to all interested parties, thus giving all interest groups an equal advantage in defending or promoting their particular causes. Under the current policy, even the frequently biased agency analyses are not widely disseminated and can be very difficult for nonagency interest groups to acquire.

Third, decision makers would be able to make their choices in the light of all available information, including public input from hearings, by mail, and through other sources. At least implicitly, and perhaps explicitly, decision makers would attach weights to the measures of attainment of the several objectives to arrive at their decisions. The decision makers would then be responsible to the various interest groups to explain their votes.

Fourth, projects and programs would be carefully monitored, with feedback on the actual attainments of each objective going both to decision makers and to the technical staff responsible for planning, design, and operation of the project. When projects perform well, both the political actors and the technical personnel who are involved with the project will be recognized and perhaps rewarded; on the other hand, deviations from designed performance can lead to penalties and improved planning procedures.

Yet there are major impediments to the adoption of such a decision setting. The greatest impediment is that many politicians do not want careful, quantitative analyses and forecasts of project impacts, at least not when such information will also be available to the public. Such information reduces their freedom to maneuver by substituting actual figures for assertions and by exposing undesirable as well as desirable effects. Political resistance to objective analysis will be greatest when the political actors see the public-sector budget as a "common property or open access resource" to be approached as a zero-sum game without regard to economic efficiency or cost-effectiveness.

Another reason for political resistance to open, quantitative project analysis is that it complicates interprogram linkages that may be part of larger strategies. President Lyndon Johnson was never greatly concerned about efficiency in the national water development program because he saw water projects as pawns in a larger strategy of coalition building for social and national defense programs. Unfortunately, neither President Johnson nor the public had valid measures of the net social costs of such projects, nor has the public any way of verifying how important such linkages really are.

Even technical personnel in various agencies have resisted the use of explicit quantitative and economic analyses for fear such tools might be used more as a weapon against them than to support their programs. For example, in fisheries and wildlife management on public lands, it seems likely today that greater social values would be generated by reducing commercial outputs at the margin and increasing recreational outputs. Such a demonstration would presumably expand the domain of those specializing in fisheries, wildlife management, and recreation. Nonetheless, there is less than full acceptance of quantitative economic analysis by fish and wildlife specialists, possibly out of fear that such analyses might be used to attack the use of the physical and biological standards that are so large a part of the decision-making apparatus of many technical personnel.

Careful monitoring and *ex post* analyses are seldom used to track the actual performance of projects. David Seckler (1984), in an excellent paper "On the Management of Public Organizations," identifies some of the reasons why monitoring and feedback are seldom used in the public sector. Seckler bases his analysis on a quotation from Peter Drucker:

"An institution which is financed by a budget ... is paid for good intentions and for 'programs.' It is paid for not alienating important constituents. ... It is misdirected by the way it is paid into defining performance as what will produce the budget rather than as what will produce contribution. ... The basic problem ... is that it is paid for promises rather than for performances." (As quoted in Seckler, 1984, pp. 140–141)

Seckler concludes that if budgets can be produced on the basis of promises, then monitoring actual performances can only cause trouble.

It is certainly true that many agencies perpetuate themselves on the basis of promised achievements, at least in the sense that the objective is to get the project built regardless of how effective the project's operation may be. States value the employment effects of project construction. And the governments of many Third World countries are so busy chasing new development projects that they ignore the maintenance and management of the projects once built.

But no agency can perpetuate itself on promises alone. The agency has to deliver something to groups that have the political power to perpetuate it. The real promise is likely to be construction work by contractors, increased employment during construction in the state, and enhanced land values around the reservoir and in the irrigated area—all without concern about the net cost to the nation. The agencies are frequently involved in tacitly promising these secondary effects to the project area at the same time they continue to pretend to the nation that the project will yield substantial national net benefits.

What can we conclude about the possibility of reform in these practices? The sine qua non of improvement in technical procedures (such as benefit–cost analysis, the setting of reimbursement rules, pricing project outputs, and so forth) is a change in political behavior and a greater concern for the performance of public projects. It also requires a greater political recognition that the public budget is not simply a common pool to be raided to get a project for one's district but a limited pool of resources, the informed use of which is intimately tied to the nation's productivity and growth as well as to its social objectives. Until this is realized by a majority of the nation's political actors, there will be no demand for objective technical analysis and consequently no advance in the more equitable divisions of the benefits and costs of water provision projects.

REFERENCES

American Rivers Conservation Council and others. 1973. *Disasters in Water Development* vol. 1 (n.p., April).

————. 1977. *Disasters in Water Development* vol. 2 (n.p., February).

Anderson, Terry L. 1983. *Water Crisis: Ending the Policy Drought* (Baltimore, Md., Johns Hopkins University Press).

Berkman, Richard L., and W. Kip Viscusi. 1973. *Damming the West* (New York, Grossman).

Burness, H. S., R. G. Cummings, W. D. Gorman, and R. R. Lansford. 1980. "United States Reclamation Policy and Indian Water Rights," *Natural Resources Journal* vol. 20 (October) pp. 807–826.

Congressional Budget Office. 1983. *Current Cost-Sharing and Financing Policies for Federal and State Water Resources Development*. Special study (Washington, D.C., July).

Cummings, R. G., D. S. Brookshire, and W. D. Schulze, eds. 1986. *Valuing Environmental Goods: An Assessment of the Contingent Valuation Method* (Totowa, N.J., Rowman and Allanheld).

Deacon, Robert T., and M. Bruce Johnson. 1985. *Forestlands: Public and Private* (Cambridge, Mass., Ballinger).

Dorfman, Robert, and Nancy S. Dorfman. 1972. *Economics of the Environment: Selected Readings* (New York, W. W. Norton).

Franklin, Douglas R., and Ronda K. Hageman. 1984. "Cost Sharing with Irrigated Agriculture: Promise *Versus* Performance," *Water Resources Research* vol. 20, no. 8 (August) pp. 1047–1051.

Haveman, Robert H. 1972. *The Economic Performance of Public Investments: An Ex Post Evaluation of Water Resources Investments* (Baltimore, Md., Johns Hopkins University Press for Resources for the Future).

Howe, Charles W., and K. William Easter. 1971. *Interbasin Transfers of Water: Economic Issues and Impacts* (Baltimore, Md., Johns Hopkins University Press for Resources for the Future).

Howe, Charles W., Joseph L. Carroll, Arthur P. Hurter, Jr., William J. Leininger, Steven G. Ramsey, Nancy L. Schwartz, Eugene Silberberg, and Robert M. Steinberg. 1969. *Inland Waterway Transportation: Studies in Public and Private Management and Investment Decisions* (Baltimore, Md., Johns Hopkins University Press for Resources for the Future).

Howe, Charles W., Dennis R. Schurmeier, and William D. Shaw, Jr. 1982. "Innovations in Water Management: An Ex-Post Analysis of the Colorado–Big Thompson Project and the Northern Colorado Water Conservancy District." Department of Economics, University of Colorado. June. Photocopy.

Hufschmidt, Maynard, Julius Margolis, and John Krutilla, with Stephen Marglin. 1961. *Standards and Criteria for Formulating and Evaluating Federal Water Resources Development*. Report of the Panel of Consultants to the Bureau of the Budget (Washington, D.C., June 30).

Kelso, Maurice M., William E. Martin, and Lawrence E. Mack. 1973. *Water Supplies and Economic Growth in an Arid Environment: An Arizona Case Study* (Tucson, University of Arizona Press).

Krutilla, John V. 1961. "Welfare Aspects of Benefit–Cost Analysis," *Journal of Political Economy* vol. 69, pp. 226–235.

————. 1966. "Is Public Intervention in Water Resources Development Conducive to Economic Efficiency?" *Natural Resources Journal* vol. 6 (January) pp. 60–75.

————. 1967. *The Columbia River Treaty: The Economics of an International River Basin Development* (Baltimore, Md., Johns Hopkins Press for Resources for the Future).

————, and Otto Eckstein. 1958. *Multiple Purpose River Development: Studies in Applied Economic Analysis* (Baltimore, Md., Johns Hopkins Press for Resources for the Future).

National Water Commission. 1973. *Water Policies for the Future* (Washington, D.C., Government Printing Office, June).

North, Ronald M., and Walter P. Neely. 1977. "A Model for Achieving Consistency for Cost-Sharing in Water Resource Programs," *Water Resources Bulletin* vol. 13, no. 5 (October) pp. 995–1005.

Seckler, David. 1984. "On the Management of Public Organizations." Department of Agricultural Economics, Colorado State University. October. Mimeo.

Shabman, Leonard. 1985. "Natural Resource Economics: Methodological Orientations and Policy Effectiveness." Paper delivered at the annual meeting of the American Agricultural Economics Association, Ames, Iowa, August.

Squire, Lyn, and Herman van der Tak. 1975. *Economic Analysis of Projects* (Baltimore, Md., Johns Hopkins University Press for The World Bank).

U.S. General Accounting Office. 1981. *Federal Charges for Irrigation Projects Reviewed Do Not Cover Costs*. PAD-81-07 (Washington, D.C., March 13).

————. 1985. *Report to the Honorable Howard M. Metzenbaum: Bureau of Reclamation's Central Utah and Central Valley Projects Repayment Arrangements*. GAO/RCED-85-158 (Washington, D.C., September 9).

Young, Robert A., and William E. Martin. 1967. "The Economics of Arizona's Water Problem." *Arizona Review* vol. 16, no. 3.

LESSONS IN
POLITICS AND ECONOMICS
FROM THE SNAIL DARTER

Robert K. Davis

The case of the snail darter,[1] or the Tellico case as it came to be called after the project at issue, has much in common with research and policy interests of John Krutilla. The case involved benefit–cost analysis of a water resources project—a tradition in which Krutilla figures so prominently (see Kneese, in this volume)—and it also concerned a comparison of the economics of replacing an irreproducible asset (specifically, a free-flowing river) with a reproducible asset (a dam and reservoir). The definition of that problem is directly traceable to the classic "Conservation Reconsidered" (Krutilla, 1967; reprinted in the appendix in this volume).

A direct descendant of the pioneering studies of Krutilla and Eckstein (1958) and Eckstein (1958), the benefit–cost analysis presented in this paper was one of many factors in an extended and sometimes heated decision-making effort that involved not only a Supreme Court decision but, ultimately, various players from all three branches of the federal government, and eventuated in a nine-month legislative battle. The source of the conflict was the expected incompatibility of a newly discovered, endangered species of fish, *Percina tanasi*—the snail darter—with the completion of a federal public works project, the Tellico Dam and Reservoir, being constructed by the Tennessee Valley Authority (TVA) on the Little Tennessee River.

The purpose of this paper is to consider the economic analysis within the context of the special set of circumstances in which it had to be carried out, and to draw such general lessons as can be drawn from that experience.

In terms of the methodology of the economic analysis discussed below, the Tellico case closely resembles that of the Hells Canyon case (Krutilla and Fisher, 1985, chaps. 5 and 6), which has become a textbook example of the application of economic analysis to natural resource protection issues. In each case, the alternatives to be compared were the preservation of a unique, free-flowing stream and the completion of a dam and reservoir. (The Tellico case did *not* involve analysis of the trade-offs between developing a water resources project and preserving an endangered species, since Congress had precluded any direct evaluation of the benefits of preserving the snail darter by requiring that the committee which was empowered to decide the case consider only the benefits of the proposed project and the benefits of any alternatives consistent with preserving the species.[2])

The results of the Hells Canyon and Tellico cases differ. Hells Canyon remains in its natural state, whereas the dam and reservoir on the Little Tennessee have been completed (although not, as it turns out, at the expense of the snail darter). In the case of Tellico there was an overpowering commitment to complete a water resources project that had in fact been planned decades earlier and that was finally under way.

Tellico brought attention to the Endangered Species Act of 1973 as prohibitive policy—a strong, comprehensive, symbolic statement—and how government can deal with such a policy. (For an illuminating discussion of the act as prohibitive policy, see Yaffee [1982]). Before the 1978 amendments were passed, the Endangered Species Act called for no discretion or balancing. If a species or its critical habitat were to be jeopardized by an action of a federal agency, that action was forbidden. The snail darter, however, shattered the prohibitive purity of the act. Tellico and a sister case, Grayrocks, which involved the whooping crane and a water project, were the first two cases to be considered by the newly established Endangered Species Committee—the committee having been created by the 1978 amendments which were specifically designed for making a choice between each water project and wildlife species involved.

After a brief introduction to Tellico and the snail darter, the following sections relate the sequence of events that eventually involved this author, then a member of the Office of Policy Analysis in the U.S. Department of the Interior, in the benefit–cost analysis of the project and its alternatives. It discusses the events and conflicting forces leading to the completion of the Tellico project and then presents some reflections on the work of the policy analyst within such a setting.

The case raises a number of worthwhile questions about the economics and politics of water projects and at the same time provides a good illustration of how Washington works. It also calls into question the usefulness of benefit–cost analysis in situations of high emotional content and obvious politics. In particular, in cases where an agency is single-mindedly pursuing a water resources project, can benefit–cost analysis be anything but a handmaiden to the agency objective?

Introduction

The Tellico Project

Completed in 1979, the Tellico project—the dam and reservoir on the Little Tennessee River a short distance from where it joins the Big Tennessee south of Knoxville in a reach known as Watts Bar Lake—had been conceived in 1936 as part of the grand design for the Tennessee Valley.[3] It was planned that a canal cut between Fort Loudon Lake (upstream from Tellico) and the new Tellico reservoir would extend navigation into the new lake and create enough head to generate additional electricity from the hydroelectric plant at Fort Loudon Dam. Wartime shortages, however, had caused postponement of the Tellico project in 1942. It was reproposed with modifications in 1963, and Congress reauthorized it and approved its initial appropriation in 1966. Construction began in 1967.

For reasons explained below, the conflict over the Tellico project and the snail darter began to take shape in about 1976 when much of the Little Tennessee above the dam site was designated as the critical habitat of that endangered species. But Tellico had first come under fire some ten years earlier when Congress was in the process of reauthorizing the project: opponents of the structure, in congressional hearings at that time, emphasized the unique natural characteristics of the river and the cultural and historic values of the valley as arguments against the dam. These values included the archaeological record left by the Cherokee and their precursors and the presence of the sites of early European occupation of Tennessee. As has been said, Congress approved the appropriation for the project in spite of arguments to the contrary.

In 1971, with project construction under way for about four years, opponents of the project filed a suit in federal court successfully contending that the Tennessee Valley Authority had not filed an adequate environmental impact statement (EIS), as required by the National Environmental Policy Act (NEPA) of 1969. (A twenty-one-month delay in construction ensued while TVA prepared the EIS, which was approved in 1973, by which time half of the project's funds had been spent.)

In preparing the EIS required by the National Environmental Policy Act, TVA emphasized the benefits of multipurpose water projects in the Tennessee Valley and asserted that the analysis of alternative means for supplying comparable benefits is basic to the analysis of a particular project (Tennessee Valley Authority, 1972). Although the act requires that an environmental impact statement address the alternatives to the project in question, TVA never seriously considered nonreservoir alternatives, according to a U.S. General Accounting Office (1977) report. The EIS considered, instead, four scaled-down versions of the dam and development of thirty-three miles of river, and rejected all of the versions because of their "failure to realize the benefits that will be provided by the Tellico project" (Tennessee Valley Authority, 1972, p. I-1-46).

The Snail Darter

As mentioned, the snail darter—*Percina tanasi*, a member of the perch family and a three-inch-long riffle fish with the unique habit of feeding on snails— assumed its role in the Tellico controversy in the mid-1970s, at which time it was believed to exist only in the Little Tennessee River. ("Tanasi," the name of an Indian village on the Little Tennessee in the 1700s, is Cherokee for "Tennessee.")

The seemingly tardy entry of the snail darter into the chronology of this story is explained in part by the fact that it was only discovered on August 12, 1973, by David Etnier, a biologist at the University of Tennessee. Two years before actually discovering the fish, Etnier had mentioned that four endangered species—the blotchside logperch, the smoky madtom, the spotfin chub, and the then-undescribed darter—might be in the Little Tennessee River and, if so, would be threatened by the Tellico Dam (Tennessee Valley Authority, 1972, p. I-1-20) because the shallows where the fish spawned would be inundated by the reservoir. At the time of its discovery, the darter was spawning in the shallows of the river above the dam. The fingerlings then moved downstream, matured in one year, and came back upstream to spawn. The full reservoir would inundate the spawning areas, making them uninhabitable by the fish. (As of 1978, the dam, although dry, was not impeding the downstream migration of the fingerlings, but it did impede the upstream migration of the mature fish, because the currents in the water passing through the concrete sluices of the dam were too strong for the fish to navigate.)

Briefly, the sequence of events after Etnier actually found the snail darter in 1973 is as follows. The U.S. Fish and Wildlife Service listed it as an endangered species in 1975. In 1976, publication of the description of *Percina tanasi* (Etnier, 1976) established the species status of the fish.[4] In the spring of the same year, the U.S. Fish and Wildlife Service listed the Little Tennessee as critical habitat for the snail darter, and in October 1976 the agency delivered its biological opinion: "The continued existence of the snail darter will be jeopardized and its critical habitat will be destroyed" if the Tellico Dam is completed and the sluice gates are closed on January 1, 1977, as planned.[5] Since the Endangered Species Act of 1973 precluded any federal action that could jeopardize an endangered species or its habitat and since the TVA was intent upon completing its project, the stage was set for conflict.

The Issue Moves to Court

The following chain of events was precipitated by a native of the Tellico area, Hiram G. Hill, Jr., who had become interested in the snail darter–Tellico issue as a student in the School of Law at the University of Tennessee. In early 1976, Hill, a student of Zygmunt J. B. Plater (who subsequently argued the case before the Supreme Court), filed suit in the U.S. District Court in Tennessee to enjoin the Tellico project as a violation of the Endangered Species Act of 1973.[6] Although the district court refused to grant an injunction that would be "an absurd result," it agreed that closure of the dam would jeopardize the snail darter. The Sixth Court of Appeals in Cincinnati then enjoined closure of the dam but permitted construction to continue as long as it did not endanger or jeopardize the snail darter.

The case then moved to the U.S. Supreme Court where it was heard in the spring of 1978. Attorney General Griffin Bell had been asked by President Carter to plead the case of the snail darter, but he had refused (Bell, 1982, p. 44). Instead, Bell appeared in court on behalf of TVA and ridiculed the situation in which a three-inch fish was holding up a TVA project. The government's case was weakened because its brief contained material from the U.S. Department of the Interior taking the side of the snail darter. The respondents in the case were represented by Zygmunt Plater.

The decision handed down on June 15, 1978, by the Supreme Court upheld the Endangered Species Act, stating that "Congress had spoken in the plainest of words making it abundantly clear that the balance has been struck in favor of affording endangered species the highest of priorities, thereby adopting a policy which it described as 'institutionalized caution' " (*Tennessee Valley Authority* v. *Hill*, 11 ERC 1721). The majority opinion made it clear that if Congress was unhappy with the outcome of the case, its solution was to change the law.

Mr. Justice Lewis F. Powell, Jr., who wrote the dissenting opinion, was bothered by the "absurd result" of the majority opinion and concluded that "Congress will amend the Endangered Species Act to prevent the grave consequences made possible by today's decision. Few if any members of that body will wish to defend an interpretation of the Act that requires the waste of at least $53 million . . . and denies the people of the Tennessee Valley area the benefits of the reservoir that Congress intended to confer" (11 ERC 1728).

Shortly after the Supreme Court decision, there was evidence of shifting attitudes in the Tennessee Valley Authority toward the Tellico project and in Congress toward the Endangered Species Act. Developments reflecting these changes included the collaboration of TVA and the U.S. Department of the Interior on a study of alternatives to Tellico, and the passage of amendments to the Endangered Species Act.

Collaborative Study by TVA and Interior

In 1977 President Carter had appointed S. David Freeman to the TVA board, replacing Aubrey J. Wagner, who as board chairman since 1962 had strongly defended Tellico. As chairman, Freeman began to question publicly the soundness of the Tellico project and to suggest that perhaps there were superior alternatives. In testimony before the House Committee on Fisheries, Wildlife, Conservation and the Environment shortly after the Supreme Court decision, Freeman suggested that it might be just as well for taxpayers that legal constraints on completing the dam would necessitate a harder look at how to best make use of the government's investment in the land, and he went on to suggest that the real waste of the taxpayers' money might be in flooding the land (Freeman, 1978).

Another indication of changing perspectives was the collaboration in the summer of 1978—after many previous failures to consult on the issue of the endangered snail darter—between TVA and the Department of the Interior on a report on the alternatives to the Tellico project (Rechichar and Fitzgerald, 1983). Economists, including this author, from the Office of Policy Analysis of the Department of the Interior participated in the benefit–cost review and, as it turned out, delivered the first substantive criticisms to TVA's benefit–cost analysis, which until then had been impregnable: TVA's claimed navigation benefits were found to be unsupportable, its recreation benefits analysis to be primitive, and its claim to regional benefits unacceptable in a national accounting framework. (The recreation analysis was reworked with help from Interior economists and made credible. Under pressure from the non-TVA economists and possibly from its own chairman, TVA began to reduce the benefits claimed for the project [Tennessee Valley Authority, 1978].)

As a result of this collaborative effort, a joint report was to have been issued by TVA and the Department of the Interior on the alternatives to the Tellico project. A draft of this report was issued on August 10, 1978, but by the time of the final report in December 1978 it was apparent that, as explained in the next section, an endangered species committee would be established to settle the Tellico case. Thus, the Department of the Interior could not take joint responsibility for the analysis and findings of the report.

The TVA report presented two distinct pieces of evidence that there was movement within that agency on the questions surrounding Tellico:[7] (1) for the first time, TVA had found an alternative—protection and use of the free-flowing river—with net benefits of comparable magnitude to those of the reservoir; and (2) benefits claimed for the dam were revised downward in response to criticism from non-TVA economists.

Endangered Species Act Amendments of 1978

After the June 1978 Supreme Court decision, Congress began to consider changes to the Endangered Species Act, and on July 19, 1978, the Senate passed amendments to the act. Briefly, the amendments came about as follows. Congress could not legislate an exemption for the Tellico project merely by continuing to appropriate funds for construction. It would have to pass legislation exempting the project from the Endangered Species Act. But John C. Culver, the Democratic senator from Iowa who headed the Resource Protection Subcommittee of the Senate Committee on Environment and Public Works, successfully argued against all attempts under the 1973 act to have Congress declare the Tellico Project exempt. He had argued that if Congress legislated an exemption for Tellico, it would have "one exemption case per week" in which it would be forced to weigh the merits of an action against the protection of endangered species, something for which Congress is ill-equipped.

Instead, his committee, which had charge of reauthorizing the Endangered Species Act, proposed a way of dealing with irreconcilable endangered species cases: the creation of a committee of cabinet officers, to be called the Endangered Species Committee, that would be empowered to decide such cases. (The amendments that created the committee came to be known as the Culver–Baker amendments—Culver eventually having been joined in this effort by Tennessee Senator Howard H. Baker, Jr.)

The Culver–Baker amendments were debated on July 18 and 19, 1978 (see Drew [1979]). Culver had opened the debate with a fervent plea for the preservation of species[8] and went on to point out that, without the flexibility afforded by the committee process, the act would be under pressure for elimination or emasculation. Culver had come to the floor with a unanimous position from his Environment and Public Works Committee, but during the debate he faced amendments from both sides that would undo the compromise. Senator John C. Stennis of Mississippi offered an amendment that vented the frustrations of those senators who saw Tellico as a symbol of development stymied. The Stennis amendment called for "grandfathering" all of the projects begun before the 1973 enactment of the Endangered Species Act and would have allowed the head of the agency to determine which projects to finish in the case of a conflict with an endangered species. The debate on the Stennis amendment laundered in public the concerns over the apparent inflexibility of the act, but it also gave Culver the opportunity to expand on the advantages of a committee "better equipped by background, training and expertise to make informed scientific, knowledgeable judgments, not to be buffeted by the political winds of the moment."[9] The Stennis amendment lost, 76 to 22.

Senator Gaylord A. Nelson of Wisconsin, on the other hand, represented the forces that wanted the act to remain unchanged. His position was that all of the conflicts except Tellico had been resolvable and that Tellico should not have been started—nor should it be completed.[10] Nelson saw the joint authorship of the amendment as suspicious and thought that the amendment "punches a big hole in a very good law. . . . [O]ur wisdom is insufficient for me to trust a handful of people to make that decision [on survival of a species] for us."[11]

At this point Senator Baker entered the debate to defend the committee process for exemption, saying that he "did not believe exemption for particular projects is a legitimate function of Congress." But he also made clear his position on Tellico: it might have been a mistake to build the dam, but "you cannot go back and undo that decision and you cannot carry off that $116 million worth of concrete. My point is you ought to go ahead, finish it and make the law conform to it."[12] The debate continued for another day, but on the afternoon of July 19, 1978, the Culver–Baker amendments came to a vote and were passed, 94 to 3.

On November 10, 1978, President Carter signed the Endangered Species Act Amendments of 1978, which created a committee to decide on irreconcilable cases involving endangered species. The law directed the Endangered Species Committee to begin proceedings within thirty days and to consider Tellico and make a decision within ninety days of the law's enactment. Failing action by the committee, Tellico would become exempt. The charge to the committee was as follows: in considering the case, it could grant an exemption "if it finds there are no reasonable and prudent alternatives to the agency action, the benefits of such action clearly outweigh the benefits of alternative courses of action consistent with conserving the species or its critical habitat, and such alternative action is in the public interest."[13]

Economic Analysis Prepared for Committee

Despite the time limits on the proceedings, it was December before the Endangered Species Committee process was set up, and mid-December before the Office of Policy Analysis in the Department of the Interior was assigned to staff the committee. Staff work began on December 22. A first draft report was prepared in two weeks. A second draft, done the next week, assimilated the record of hearings held in Knoxville and Washington, D.C. (on January 8) and other submissions and was circulated to committee members for comment on January 12, 1979. The final report was prepared on January 19, four days before the committee meeting.

The Office of Policy Analysis staff members assigned to work for the committee used the legislative history (U.S. House of Representatives, 1978) to understand crucial terms and concepts. Although the legislative history allowed some leeway in the choice of procedures for establishing the benefits of various alternatives, the legislative language was influenced by the instructions pertaining to impact analysis contained in Executive Order 11949. The staff, however, chose *Principles and Standards for Planning Water and Related Land Resources* of the U.S. Water Resources Council (1979) as its standard for the economic analysis.

By law the committee had the following options: it could (1) deny an exemption, (2) grant an unqualified exemption, or (3) grant an exemption while requiring one or more mitigation measures (U.S. Department of the Interior, 1979, p. 5.2), such as delaying the closure of the Tellico Dam long enough to be sure that the populations of snail darters in other rivers were going to survive—by this time the fish had been transplanted in two other rivers; finding other rivers for transplantation; or requiring studies of propagation in captivity.

Benefits of the Project

Table 1 summarizes the benefits and costs of the project and of the river alternative. The committee staff estimated the total annual project benefits to be $6.50 million. Originally, TVA had estimated that total benefits would be $16.53 million, distributed among land enhancement, flood control, navigation, power, recreation, water supply, and employment purposes as shown in column (1) of table 1. With costs at $5.02 million and a benefit-cost ratio of 3.29, the project appeared to provide a comfortable margin of net benefits. But after the criticism of its results mentioned earlier, TVA revised its benefit estimates downward by a considerable margin, as shown in column (2). For one thing, TVA could not count land enhancement benefits and also navigation and recreation benefits, both of which are free services that are likely to be capitalized in land values (Knetsch, 1964). The change in the flood control estimate (see column (2)) took place because TVA began using an altered concept of the maximum probable flood. In addition, TVA's navigation benefits fell by almost two-thirds under criticism from a variety of sources; the final figure still represented what some felt was a highly dubious claim.

On the other hand, power benefits, as shown in the table, reflected an increase in the cost savings that resulted from substituting additional hydro generation for the nuclear and coal-fired plants that would have been operated instead. The methodology for estimating the recreation benefits was also improved in that the estimate was based on the travel-cost/demand method and incorporated the concept of substitution between sites. (Recreational visits to the Tellico reservoir that would otherwise have occurred on another reservoir were not counted.[14]) Employment benefits were a victim of the classic arguments over secondary benefits; water supply benefits applied only to one small town, Vonore, Tennessee, which later would refuse to accept them.

In TVA's 1978 estimate, costs also diminished because funds had already been sunk in the project. TVA had estimated the remaining costs for the dam project to be $35.1 million, which included $14.5 million to enable spillways to handle a larger maximum flood. The total benefits of $6.85 million less the incremental costs of $3.19 million left net benefits of $3.66 million, or a benefit–cost ratio of 2.15, in 1978.

The committee staff estimates, in column (3) of table 1, show a change of major importance on the cost side: the staff assigned an opportunity cost to the project lands. The reasoning behind this change was that the project still faced an opportunity cost for the land because, although it might have been purchased, it was not irretrievably committed until the reservoir was created.[15]

Table 1. Estimated Annual Benefits and Costs of the Tellico Dam and Reservoir and the River Development Alternative, Ignoring Environmental Factors (millions of 1978 dollars)

Benefits and costs	Original TVA estimate[a] (1)	1978 Revised TVA estimate[b] (2)	Committee staff estimate[b] (3)	River development alternative: committee staff estimate[b] (4)
Total benefits (B)	16.53	6.85	6.50	5.10
Land enhancement[c]	1.62	0.34	—	—
Flood control	1.13	1.04	1.04	—
Navigation	0.89	0.31	0.10	—
Power	0.89	2.70	2.70	—
Recreation	3.70	2.30	2.50	3.10
Water supply	0.16	0.05	0.05	—

Agriculture[d]	—	0.11	0.11	2.00
Employment	8.14	—	—	—
Total costs, $(r + d)K$	5.02	3.19	7.22	6.29
Dam	5.02	3.19	3.19	2.26
Land[e]	—	—	4.03	4.03
Net benefits, $B - (r + d)K$	11.51	3.66	−0.72	−1.19
Benefit–cost ratio, $B/(r + d)K$	3.29	2.15	0.90	0.81

Note: Dash = not applicable.

Source: Edward M. Gramlich, *Benefit–Cost Analysis of Government Programs*, © 1981, p. 150. Adapted by permission of Prentice-Hall, Inc., Englewood Cliffs, N.J.

[a] Taken from the U.S. General Accounting Office, *The Tennessee Valley Authority's Tellico Dam Project: Costs, Alternatives, and Benefits* (Washington, D.C., October 1977) table 1. All numbers are converted to 1978 dollars by multiplying by 2.23, the ratio of the gross national product (GNP) price deflator in 1978 to that in 1968 (the base year).

[b] Taken from U.S. Department of the Interior, *Tellico Dam and Reservoir*, Staff Report to the Endangered Species Committee (Washington, D.C., 1979, exhibit 3). Midpoints are used whenever ranges are shown in the original source.

[c] For land surrounding the reservoir.

[d] On land included in and surrounding the reservoir.

[e] Of land included in the reservoir.

The staff used current market values of $2,500, $1,400, and $650 per acre, respectively, for "prime," "statewide-important," and "other" classes of land for a composite value of $1,196 per acre and a total value of $43.2 million.[16]

The cost of the land was annualized at the private opportunity cost of capital: the 10 percent discount rate. Had the staff used the 6.63 percent discount rate on the land value, its estimate of net benefits for completing the project would have been a positive $0.55 million instead of a negative $0.72 million. The difference in net benefits on this item made the percentage a crucial issue. We believed, however, that using the private discount rate to annualize the cost of land acquired at prices set in private markets was the correct procedure (see Eckstein [1958, p. 146]).

Benefits of Alternative Courses of Action

Column (4) in table 1 shows the benefits and costs of the river development alternative. The principal benefits of the alternative come from recreation and agriculture; the opportunity costs of land are the same as in the dam–reservoir alternative. Additional costs were involved in river development, however, in the removal of the dam, restoration of historic sites, and construction of highways. Thus, the estimated costs of river development exceeded the benefits by a margin of $1.19 million, and the river development alternative was inferior (by about $0.5 million per annum) to that of completing the project.

TVA's attempt to find agricultural benefits in the river development alternative led it to envision 1,000 acres of high-value fruit and vegetables growing on seventy-three farms and, additionally, sixty dairy farms. TVA was prepared to go into partnership with farmers using intensive agricultural practices to take advantage of the markets for seasonal fresh fruits and vegetables in Chattanooga, Knoxville, and Atlanta. The 1979 edition of the U.S. Water Resources Council's *Principles and Standards* did not allow specialty crops nor livestock benefits to be used to account for agricultural benefits, however. To be consistent, the staff cut TVA's agricultural benefits to $1.5 million but allowed $0.5 million in resource cost savings for the employment of under-employed resources. Agricultural benefits under the river development alternative thus totaled $2.0 million (U.S. Department of the Interior, 1979, p. 2-3).

The Special Case of River Recreation

In its 1978 report, TVA used unit day values to estimate recreation days. Moreover, its model for estimating recreation use on a new site did not correct for use transferred in from existing sites. By December 1978, TVA's recreation estimates were based on a travel-cost model, its estimates of use took account of alternative sites, and its comparisons of alternatives were sensitive to the comparative uniqueness of reservoir and river recreation. Table 2 shows the basis for the recreation benefit calculations. The cross-elasticity of demand, which is used to account for substitution, is assumed to be between 0.5 and 5.0 for the river alternative; the cross-elasticity of the reservoir is assumed to be in the range of 2.0 and 20.0. Using 0.5 for the river cross-elasticity and 10.0 for the reservoir cross-elasticity, allowing for slightly higher growth rates in the demand for recreation on the river, and assuming (for lack of data) equal capacity constraints, recreation benefits are $3.10 million for river development and $2.50 million for reservoir development.[17]

The Committee's Decision

When the Endangered Species Committee convened on January 23, 1979, the Tellico case opened with a ten-minute briefing on the staff report. A question from the Secretary of the Army established the fact that TVA had not made a recommendation to the committee. The representative of the state of Tennessee, William R. Willis, Jr., of Nashville, confirmed that TVA's river development alternative was a reasonable alternative to the completion of the dam. Then, with almost no further preliminaries, Charles Schultze, chairman of the President's Council of Economic Advisors, asked to be recognized.[18] Upon being given the floor by Chairman Cecil Andrus, Schultze said:

Well, somebody has to start. . . . I have not prepared a resolution; however, I think the sense of it would be clear. It seems to me the examination of the staff report (which I thought was excellently done) would indicate that it is very difficult . . . to say there are no reasonable and prudent alternatives to the project. The interesting phenomenon is that [this] project . . . is 95 percent complete, and if one takes just the cost of finishing it against the benefits and does it properly it doesn't pay, which says something about the original design.

Schultze went on:

It's also true that the particular river development plan posed by TVA as an alternative also has negative net benefits, slightly larger, negative benefits. I note that the staff report points out the market value of the raw land involved, which is still available for liquidation as an alternative, is something in the neighborhood of $40 million which appropriately discounted gives you 4 million dollars a year. The staff notes that in further developing any specific river development plan, the TVA would have to look very carefully at what mix of private and public ownership, lease and purchase would maximize the total value. On the basis of this, it seems to me that a completion of the project returning negative net benefits, that a development alternative, which at the moment also has negative benefits, but only slightly larger, an alternative which does preserve some archaeological sites, some scenic value, I don't see how it is possible to find that there is no reasonable and prudent alternative, nor do I see how it is possible to find that the benefits of completing the project clearly outweigh the benefits of alternatives consistent with conserving a species. Therefore, Mr. Chairman, . . . I would move that we deny an exemption. (U.S. Department of the Interior, Endangered Species Committee, 1979, pp. 26–27)

Table 2. Annualized Recreation Benefits for the Reservoir and River Development Alternatives (millions of dollars)

Cross-elasticity[a]	Capacity assumption[b]	Reservoir			River		
		2.5%	3.5%	5.0%	3.5%	5.0%	7.0%
0.5	L				2.409	2.829	3.390
	M				2.558	3.040	3.732
	H				2.673	3.300	4.241
1.0	L				2.305	2.673	3.163
	M				2.495	2.924	3.538
	H				2.651	3.244	4.109
2.0	L	1.998	2.201	2.554	2.232	2.546	2.955
	M	2.169	2.447	2.867	2.451	2.826	3.356
	H	2.264	2.658	3.278	2.639	3.193	3.977
5.0	L	1.966	2.139	2.436	2.176	2.441	2.766
	M	2.154	2.408	2.773	2.416	2.742	2.186
	H	2.262	2.646	3.226	2.629	3.146	2.847

10.0	L	1.954	2.115	2.389	2.155	2.399	2.686
	M	2.148	2.393	2.735	2.402	2.708	3.112
	H	2.262	2.642	3.204	2.625	3.126	3.788
20.0	L	1.948	2.102	2.363			
	M	2.145	2.384	2.713			
	H	2.262	2.639	3.191			
30.0	L	1.946	2.098	2.354			
	M	2.144	2.382	2.707			
	H	2.262	2.638	3.187			

Source: Adapted, by permission of Butterworth Scientific Limited, from F. Reed Johnson, "Federal Project Evaluation and Intangible Resources," *Resources Policy* (September 1981) p. 205.

[a] Cross-elasticity is the ratio of the percentage change in visitation to the percentage change in the price of alternative recreation opportunities. This coefficient is a measure of the availability of close substitutes for recreation at the site.

[b] L = design capacity; M = 1.30 × design capacity; H = 2.00 × design capacity. Design capacity is defined as boating and camping activities only.

The motion was seconded and adopted. The snail darter had prevailed unanimously over the dam. Chairman Andrus closed the proceedings with comments about how well the committee process had worked on this occasion and how that boded well for the future.

The Road to Exemption

The day after the Endangered Species Committee made its decision, Senator Baker was quoted in the January 24, 1979, edition of the *Washington Post* as saying: "If that's all the good the committee process can do, to put us right back where we started from, we might as well save the time and expense. I will introduce legislation to abolish the committee and exempt the Tellico Dam from the provisions of the act." Baker's statement was the beginning of a nine-month fight by the Tennessee congressional delegation to overrule the committee. At a hearing of the Culver committee (i.e., the Resource Protection Subcommittee of the Senate Committee on Environment and Public Works) in May 1979, Baker repeated his disavowal of the Endangered Species Committee's action despite his earlier support for the amendment creating the committee.[19] He also introduced an amendment to exempt the Tellico project in order to "close the gates on a dam we bought and paid for." The committee voted against the Baker amendment, however, 10 to 3.

Yet Baker continued to pursue the exemption. The Culver committee had been working on an extension of the Endangered Species Act, which came to the floor of the Senate on June 13, 1979, and Baker offered his amendment for exemption. The amendment failed by a vote of 52 to 43, and this step exhausted the normal legislative remedies available to the Tennessee delegation.

By June 18, 1979, the action had moved to the House. When the House public works bill containing energy and water appropriations was being considered late in the afternoon, Tennessee Congressman John J. Duncan proposed an amendment to exempt Tellico from any federal law impeding its completion. Duncan asked that the amendment not be read and instead began to explain it. Before he could finish, John T. Myers of Indiana jumped to his feet and said that he had read the amendment and would accept it. Tom Bevill of Alabama, the chairman of the Appropriations Subcommittee that wrote the bill and in whose district the Tennessee–Tombigbee waterway was under construction, announced that he also accepted the Duncan amendment. Thus, with no further explanation of its contents, the amendment was passed by what a UPI wire service story of June 18 called a "mumbled voice vote apparently with few on the House floor aware of what was happening."

The amendment authorized and directed TVA to complete the dam. It was the Tennessee delegation's first victory over the snail darter in many months. The Senate Appropriations Committee then approved the House version of the bill, but on July 18 Culver again prevailed on the floor of the Senate (by a 53-to-45 vote) and the Tellico language was removed from the Senate appropriations bill. The language was reinserted by the conference committee, however, and on September 10, 1979, after a lengthy debate the Senate voted 48 to 44 to grant an exemption. Howard Baker thus had finally won a victory over the fish he had been calling "the bane of my existence, the nemesis of my golden years, the bold perverter of the Endangered Species Act." Although professing to have nothing personal against the snail darter—"he seems to be quite a nice little fish as fish go"—Baker said the snail darter had become the unfortunate symbol of a type of environmental extremism that could spell the doom of the environmental protection movement and that advocated the perversion of the Endangered Species Act as a device "to challenge any and all Federal projects."[20]

The Rhetoric of Exemption

The rhetoric used to argue for the exemption of the Tellico project drew a great deal of strength from the U.S. energy situation.[21] In 1979 the nation was feeling the inflationary bite of the price hikes of the Organization of Petroleum Exporting Countries (OPEC), and President Carter was promoting energy independence. When completed, Tellico would light 20,000 homes with hydroelectric power. Senator Baker thus urged his colleagues "to seize this opportunity to redeem our commitment to energy production." The Tellico supporters had also found openings for disputing the analysis of benefits and costs presented to the Endangered Species Committee. Senator James R. Sasser of Tennessee switched from Baker's earlier tactic of declaring the matter to be an environmental question; rather, Sasser said, "It is an economic question," and he proceeded to argue against letting the $111 million already sunk in the project go down the drain. (The $111 million claimed as the sunk cost of the dam was repeatedly refuted by the Culver forces, who reiterated that only $22 million had actually been spent on the dam itself.) Sasser also cited the earlier benefit–cost ratio of 2.3 found in TVA's alternatives report (Tennessee Valley Authority, 1978) and accused the committee of making its decision based on some "creative accounting." Senator J. Bennett Johnston, Jr., of Louisiana continued the attack on the committee's economics, in particular, objecting to the inclusion of the opportunity costs of the land among the costs of completing the project.

Although the exemption forces were about to win, Senator Culver summarized the list of arguments against the motion. They included

♦ the difficult responsibility that Congress would now face in having to make highly complex decisions on more cases of endangered species;
♦ the Tellico project's lack of economic viability and the Endangered Species Committee's finding that the completed project would not pay;
♦ the waste of most of the prime farmland in the project area (it would be flooded by the reservoir, according to the U.S. Department of Agriculture);

+ opposition to completion of the project by the Office of Management and Budget;

+ the minute (one-thousandth) portion of TVA's total energy capacity that Tellico would provide, in addition to which TVA was deferring further construction of nuclear power units;

+ the expenditure of only $22.5 million (rather than $111 million, as claimed by project supporters) on the dam by that time and the fact that all of the rest of the federal expenditures could be put to beneficial use;

+ the endangered status of the snail darter and the question of the viability of transplanted populations;

+ and the violation of the jurisdictional prerogatives of the Environment and Public Works Committee that exemption would constitute; and, finally,

+ the unprecedented exemption of a project from all other laws.

Why the Committee Was Snubbed

The 48-to-44 vote on September 10, 1979, in favor of exempting the Tellico project was a reversal of the 53 to 45 margin of July 17 because three northern senators (Mathias, Ribicoff, and Danforth) and three western senators (Cannon, Dole, and Gravel) switched to the exemption side on the final vote. The reversal also occurred because eight senators who had previously supported Culver's position were not present to vote.[22] In addition, there were four senators (Cohen, Magnuson, Chiles, and Stone) who switched from favoring exemption to opposing it.

In all three of the previous votes, the division of the Senate on Tellico had followed regional lines. The South had supported the water project, and the rest of the country had opposed it. In the final vote, the West swung to the other side and favored exemption. Table 3 shows the breakdown of the voting by region.

Table 3. Regional Breakdown of Senate Vote on Tellico Exemption

	July 17, 1979		September 10, 1979	
Region	No. for exemption	No. against exemption	No. for exemption	No. against exemption
North	7	30	9	25
South	23	5	21	6
West	15	18	18	13
Total	45	53	48	44

Source: *Congressional Quarterly Almanac 1979* (Washington, D.C., Congressional Quarterly, pp. 31-S, 45-S).

A number of lessons in political reality can be drawn from this campaign. One is found in the words of Howard Baker: "This project has been bought and paid for." In other words, a deal that has been made must be kept; a project that has been authorized must not be deauthorized. Once such a deal is made and a project is authorized and begun, its beneficiaries have established claims—property rights—to the promised benefits.

The second and corollary proposition is that sunk costs cannot be ignored because they are the down payment on the "contract." To ignore sunk costs is to disavow the bargain that has been struck. To the members of Congress defending Tellico, there was no consideration of the proposition that they might be "throwing good money after bad." To admit that taxpayers' money is wasted is a greater crime than the wasting of more money.

A third lesson is that when regional projects reach the final vote, they must always be presented as national projects. Although the trade-off in this case involved gains for the regional beneficiaries of the project balanced against net costs to the nation at large, little was heard of the regional arguments in the final debates.

Carter's Dilemma

The Fiscal Year 1980 energy and water appropriations bill that emerged from the Congress contained the Tellico exemption, funds for completing the dam, and funds for the Hart Senate Office Building. To make a threatened veto all but impossible, Congressman Bevill of Alabama tacked the $10.8-billion appropriations bill onto the continuing resolution to keep the government operating after October 1. There were some signs that the president might veto this exemption as he was being urged to do by Secretary of the Interior Andrus. The press began to take sides after it became apparent that Congress was going to uphold the exemption: numerous editorials called the project a waste, urged an end to the pork barrel, and generally echoed Charles Schultze's sentiments about a project that is 95 percent complete and still does not pay its way.

President Carter had established a record as an environmentalist and also as an enemy of pork barrel projects. He had vetoed the public works bill in October 1978 and recalled later that "the battle left deep scars" (Carter, 1982). In 1979, as he faced a decision on the Tellico project, he was already deeply embroiled in the issue of Panama Canal legislation, Salt II, the creation of a department of education, and initiatives in energy policy, and he was anticipating his reelection campaign of 1980. It was not a good time to pick a fight, especially one that involved letting a three-inch fish stop the federal government in its tracks. The snail darter had already borne enough ridicule for stopping a $100-million reservoir project. On September 25, 1979, the president made this statement: "It is with mixed emotions that I sign H.R. 4388, the Energy and Water Development Appropriations Bill,"[23] explaining his reason for not vetoing the bill:

While I believe firmly in the principles of the Endangered Species Act and will enforce it vigorously, I do not consider that the action by Congress on the Tellico matter implies Congressional intent to overturn the general decision process for resolving conflicts under that Act. Furthermore, I am convinced that this resolution of the Tellico matter will help assure the passage of the Endangered Species Act reauthorization, without weakening amendments or further exemptions (Office of the White House Press Secretary, "Statement by the President," September 25, 1979).

Carter also expressed the hope that by being reasonable on this bill he would get support from Congress for a water project review function to be lodged with the Water Resources Council, which itself was having difficulty getting funded (ibid.). Environmentalists were not pacified, however. Brent Blackwelder of the Environmental Policy Center said Carter "had a chance to show leadership and he blew it" (Congressional Quarterly, 1979).

TVA Wins—and Loses

By the day after the president signed the appropriations bill, TVA's bulldozers were rolling. By November 13 the last two farmers had been evicted, and at 11:23 A.M. on November 29, 1979, the gates of the dam were dropped into place. Former TVA chairman Aubrey Wagner attended the event. Wagner, according to whom the "only appropriate course of action had been closure of the gates of Tellico" (Rechichar and Fitzgerald, 1983, p. 52), told reporters, "I'm glad to see it filled, finally" (AP news wire story, November 30, 1979). The lesson to be learned on that November day was that persistence is rewarded. TVA had begun with a mission to improve navigation in the Tennessee River and to control destructive floods in the Tennessee and Mississippi River basins. It had been charged by President Franklin D. Roosevelt with the broader duty of "planning for proper use, conservation and development of natural resources of the Tennessee River drainage basin and its adjoining territory" (Hodge, 1938, p. 36). As a result of that charge, TVA had become irrevocably committed to completing a system of dams conceived in the 1930s to regulate the flow of the Tennessee and its tributaries.

Although local opposition and unprecedented regulation of its activities by the National Environmental Policy Act and the Endangered Species Act severely tested TVA's resolve, the old coalition of businessmen, newspaper editors, and local officials never wavered in its support of Tellico. Although Tennessee Congressman John J. Duncan found that, in his district of 190,000 households, 82 percent of 13,046 persons responding to a mail survey favored completion of Tellico (U.S. Department of the Interior, 1979, p. 3.1), we may never know whether or not the majority of local residents perceived a need for the dam.

Where Is the Snail Darter Now?

At the time the Endangered Species Committee met, the snail darter was known to exist in two places in Tennessee and to be established in only one of them. It now appears that the entire controversy may have revolved around a dam that was not worth building in the first place and a fish that did not really need to be saved. On August 6, 1984, the Fish and Wildlife Service downgraded the status of the snail darter from "endangered" to "threatened."

Within six years after the dam was closed, the snail darter had been found in Chickamauga Creek in Chattanooga and in Watts Bar Lake on the Big Tennessee below Tellico, a puzzling development because at the time of the committee deliberations, one would have thought that if there were any other snail darter populations in Tennessee, TVA would have found them.[24] In Watts Bar Lake, adult snail darters have been found in depths of twenty feet or more, where, according to biologists, visibility is poor, seining is costly, and only one snail darter may be the reward for a day's work. In 1981 and 1982, biologists again found darters in Watts Bar Lake but could make no positive identification of young for either of those years. It was known that if the darters could reproduce it would indicate that they could survive in deeper water as long as there were clean riffle conditions, which apparently existed in this location.

There are several other locations in the Tennessee River system—Sewee Creek, Sequatchie River, and Point Rock River—in which populations of snail darters are being found under similar circumstances, and there are also other sites in Alabama and Tennessee that are considered promising.

Conclusion

The snail darter case closely resembles that of Hells Canyon (Krutilla and Fisher, 1985). Like the Low Mountain Sheep–Pleasant Valley Dam project in Hells Canyon, the Tellico Dam was found to have a negative net benefit without even counting any environmental effects, which for Tellico included flooding one of the few remaining natural river reaches in the Tennessee Valley. In such a circumstance, Krutilla reminds us that "we could counsel against development on economic grounds without requiring analysis of the environmental costs" (Krutilla and Fisher, 1985, p. 122), the position taken on Tellico by Charles Schultze, the economist on the Endangered Species Committee.

But as in the Hells Canyon case, there was another aspect of the problem that required an analysis of the environmental amenity benefits that would have to be forgone. In Hells Canyon, it was the High Mountain Sheep Dam, which had a positive net present value of development benefits, necessitating an evaluation of the benefits associated with preserving the area in its natural condition. In the Tellico case, it was the law of exemption, which required a comparison of the project with alternatives that were consistent with the preservation of an endangered species; this meant the dam had to be compared with preserving the free-flowing Little Tennessee River.

In Hells Canyon, the pathbreaking analysis of the loss of amenities that would occur if High Mountain Sheep Dam were constructed led to the reasonable conclusion that the small gain in the net value of hydroelectricity was not worth the loss of the amenity benefits the area provided in its preserved state. In the Tellico case, the amenities preserved with the free-flowing river were sufficiently valuable that the Endangered Species Committee could say, without direct analysis of the economic values of the snail darter, that preservation was a reasonable alternative to the dam.

In their impact on policy, the cases diverge sharply. Consistent with the economic analysis, Hells Canyon was added to the National Wilderness Preservation System in 1975, and thermal power generating capacity appears to be providing an economic substitute for the hydroelectric potential of the site (Krutilla and Fisher, 1985, p. 143). Contrary to the economic analysis, Tellico Dam was completed in 1979, adding yet another body of flat water to the Tennessee Valley and ostensibly jeopardizing the existence of the snail darter.

One of the risks in pursuing applied welfare analysis is that occasionally events turn out contrary to the results of the analysis. When this happens, we must ask whether the events as they turned out have nothing to do with the analysis or whether there is a flaw in the model—the progressive model— that most analysts follow when they engage in such work. The progressive model emphasizes good science, systematic and rigorous analysis, and rational decision making based on that analysis. Progressivism is so closely identified with good behavior in government that, given a sound, systematic staff analysis in the Krutilla tradition and an Endangered Species Committee that played according to progressive rules, Congress might not have been able to vote against the snail darter without the help of sleight-of-hand legislative strategy driven by long-established claims for the project.

We do not expect as many of the actors to fail to live up to expectations as they did in the snail darter case. A president who lived the progressive faith was immobilized amidst a crescendo of opposition to the economic inefficiencies of the pork barrel and "blew it," according to one environmentalist. TVA, a blend of engineering, economic boosterism, and conservationist zeal founded on New Deal progressivism, felt sunk costs were not sunk. Finally, the scientists and technicians who were responsible for upholding the tradition of thorough biological investigation failed to find other populations of snail darters when such information might have rendered the question of preservation moot and kept the economics of Tellico from becoming a nationwide issue.

The progressive model does not help us explain such outcomes, nor does it help us to tailor our analyses to anticipate these variables. The case makes us ask whether, to be more effective in the arena of public decisions, economists must take more account of the distributional issues and existing political claims in each case. We must also ask whether economists might not be more effective advocates of economists' solutions to issues of policy. In the case of the snail darter, it is fair to question whether the staff analysis led the committee down the wrong path; and whether a more knowing analysis could have led to a committee decision that would have been both responsive to the Endangered Species Act and acceptable to the forces that were bound to have their way in the end.

It is an encouraging sign that a number of scholars are beginning to pursue these questions. On the one hand, we have analysis by economists of the behavior of economists in the policy process (Leman and Nelson, 1981; Nelson, 1987; Shabman, 1983). On the other hand, attention is also being paid to the art of persuasion in economics (McCloskey, 1985). Those of us who would continue to labor on the policy issues would do well to maintain an ongoing inquiry into the reasons for our successes and failures.

NOTES

1. The story related in this paper is based on my experiences in the late 1970s, first as the principal staff member assigned by the U.S. Department of the Interior to assist the Endangered Species Committee in the case of the snail darter and then as a close observer of the ensuing legislative battle. The serious time constraints imposed on the committee by law resulted in a schedule that allowed less than a month for the staff work that would be the basis for the decision in the snail darter case. The report produced for the committee (U.S. Department of the Interior, 1979) provides much of the substance of this paper.

2. Of course, valuation of the species would be implicit in the comparison of the project with the best of the alternatives consistent with preserving the species.

3. The history of Tellico is reviewed in the staff report to the Endangered Species Committee (U.S. Department of the Interior, 1979) and in a report of the Tennessee Valley Authority (1978).

4. Eugene Kinkead chronicles the discovery and further studies of the snail darter in *The New Yorker* (January 8, 1979, pp. 52–55).

5. The opinion was contained in a letter of October 12, 1976, from Lynn A. Greenwalt, director of the U.S. Fish and Wildlife Service, to Lynn Seeber, general manager of the Tennessee Valley Authority.

6. Most of the information in this section is taken from *Tennessee Valley Authority v. Hill*, U.S. Supreme Court decision no. 76-1701 of June 15, 1978.

7. Rechichar and Fitzgerald (1983, pp. 52–57) note that Freeman's entrance disrupted the TVA board's unanimity on the project.

8. 124 *Cong. Rec.* S10973, July 18, 1978.

9. 124 *Cong. Rec.* S11022, July 19, 1978.

10. Nelson had privately told Culver that he did not think Senator Baker would get his dam from these amendments (Drew, 1979, p. 49).

11. 124 *Cong. Rec.* S11028, July 18, 1978.

12. 124 *Cong. Rec.* S11029-30, July 18, 1978.

13. Endangered Species Act Amendments of 1978, P.L. 95-632, 92 Stat. 3758, amending 16 U.S.C. 1536, Sec. 10(i)(1) and Sec. 7(h)(1)(A)(i),(ii), as amended.

14. This argument assumes that the new and the displaced sites provide comparable services so there is no improvement in quality at the new site or avoidance of congestion at the site experiencing the displacement.

15. Letter dated January 8, 1979, from Leonard Shabman to Secretary of the Interior Cecil D. Andrus.

16. A crucial line was omitted from the staff report, however, creating the erroneous impression that a market price of $2,500 per acre had been used as the value of all the land in the reservoir site.

17. Note the parallels with the Cicchetti–Krutilla analysis of High Mountain Sheep dam in Hells Canyon (Krutilla and Fisher, 1985, chap. 6).

18. The unexpected presence of Schultze on an Endangered Species Committee finds explanation in the legislative history of the committee. The chairman of the Council on Environmental Quality was first designated as a member of this committee, together with the Secretary of the Interior, of Agriculture, and of the Army, and the heads of the National Oceanic and Atmospheric Administration (NOAA) and of the Environmental Protection Agency (EPA). The balance of the committee looked as though it might favor preservation, however, and so the president's economic advisor was substituted for his environmental advisor. Interestingly, the economist proceeded to strike a blow for the darter rather than for the dam.

19. As recorded in personal notes taken by this author at the May 10, 1979, hearing, Senator Pete V. Domenici of New Mexico expressed interest in Baker's statement that the species was not endangered. With the help of staff, Culver explained that the snail darter recovery team believed that up to fifteen more years were needed to determine the viability of the population of darters in the Hiwassee River. When Domenici declared a lack of interest in this level of detail, Culver jumped at the opportunity to make his often-repeated point that this was just the kind of detail the senate *must* review if it is to make exemptions to the Endangered Species Act in each case of irreconcilable conflict.

20. 125 *Cong. Rec.* S12274, September 10, 1979.

21. All quotations in this subsection, The Rhetoric of Exemption, are taken from the September 10, 1979, edition of the *Congressional Record* (pp. S12274-S12279).

22. The eight "no-shows" were Bayh, Durenberger, Muskie, Pell, Bumpers, Armstrong, Inouye, and Pressler.

23. According to a conversation of the author with Zygmunt Plater, the law professor who had taken the snail darter to the Supreme Court, President Carter telephoned Plater the night before he signed the bill and spoke of his mixed feelings about the act and the conflict between his sympathies for the cause of the fish and the realities of his situation.

24. Chickamauga Creek was a polluted, urban waterway criss-crossed with freeways; it was difficult to get to and unpleasant to work in. Biological surveys bypassed it for many years. When it was finally sampled, it yielded both snail darters and another endangered species, the logperch.

REFERENCES

Bell, Griffin B. 1982. *Taking Care of the Law* (New York, William Morrow).

Carter, Jimmy. 1982. *Keeping the Faith—Memoirs of a President* (New York, Benham Books).

Congressional Quarterly. 1979. September 29, p. 2140.

Drew, Elizabeth. 1979. *Senator* (New York, Simon and Schuster).

Eckstein, Otto. 1958. *Water Resources Development: The Economics of Project Evaluation* (Cambridge, Mass., Harvard University Press).

Etnier, D. A. 1976. *"Percina (Inostoma) tanasi,* a New Percid Fish from the Little Tennessee River, Tennessee," *Proceedings of the Biological Society of Washington* vol. 88, pp. 469–488.

Freeman, S. David. 1978. Testimony before the House Committee on Fisheries, Wildlife, Conservation and the Environment, 95th Cong., 2d sess., June 23.

Hodge, Clarence Lewis. 1938. *The Tennessee Valley Authority: A National Experiment in Regionalism* (New York, Russell and Russell).

Knetsch, Jack L. 1964. "Economics of Including Recreation as a Purpose of Water Resources Projects," *Journal of Farm Economics* (December) pp. 1148–1157.

Krutilla, John V. 1967. "Conservation Reconsidered," *American Economic Review* vol. 57 (September), pp. 777–786 (Resources for the Future Reprint 67).

————, and Anthony C. Fisher. 1985. *The Economics of Natural Environments: Studies in the Valuation of Commodity and Amenity Resources* Rev. ed. (Washington, D.C., Resources for the Future).

————, and Otto Eckstein. 1958. *Multiple Purpose River Development: Studies in Applied Economic Analysis* (Baltimore, Md., Johns Hopkins Press for Resources for the Future).

Leman, Christopher K., and Robert H. Nelson. 1981. "Ten Commandments for Policy Economists," *Journal of Policy Analysis and Management* vol. 1, no. 1, pp. 97–119 (Resources for the Future Reprint no. 198).

McCloskey, Donald N. 1985. *The Rhetoric of Economics* (Madison, University of Wisconsin Press).

Nelson, Robert H. 1987. "The Economics Profession and the Making of Public Policy," *Journal of Economics Literature* vol. 25, no. 1, pp. 49–91.

Plater, Zygmunt J. B. 1982. "Reflected in a River: Agency Accountability and the TVA Tellico Dam Case," *Tennessee Law Review* vol. 49, pp. 747–787.

Rechichar, Steven J., and Michael R. Fitzgerald. 1983. *The Consequences of Administrative Decision: TVA's Economic Development Mission and Inter-governmental Regulation* (Knoxville, Bureau of Public Administration, University of Tennessee).

Shabman, Leonard. 1983. "Nonmarket Valuation and Public Policy: Historical Lessons and New Direction," pp. 62–99 in John R. Stoll, Robert N. Shulstad, and Webb M. Smathers, Jr., eds., *Nonmarket Valuation: Current Status and Future Directions*, Proceedings of a Regional Workshop, May (Southern Rural Development Center and Farm Foundation).

Tennessee Valley Authority. 1972. *Environmental Statement, Tellico Project* (Chattanooga, Tenn., TVA, Office of Health and Environmental Science, February 10).

————. 1978. *Alternatives for Completing the Tellico Project* (Knoxville, Tenn.)

U.S. Department of the Interior. 1979. *Tellico Dam and Reservoir.* Staff Report to the Endangered Species Committee (Washington, D.C.).

U.S. Department of the Interior, Endangered Species Committee. 1979. "Transcript of the Meeting of January 23" (Washington, D.C.).

U.S. General Accounting Office. 1977. *The Tennessee Valley Authority's Tellico Dam Project—Costs, Alternatives and Benefits.* Report to Congress (Washington, D.C., October 14).

U.S. House of Representatives. 1978. *Conference Report on the Endangered Species Act Amendments of 1978.* H.R. Rep. No. 95-1804. 95th Cong., 2d sess.

U.S. Water Resources Council. 1979. *Principles and Standards for Planning Water and Related Land Resources. Federal Register* vol. 44, no. 242 (Dec. 14) 72,892–72,990.

Yaffee, Steven L. 1982. *Prohibitive Policy: Implementing the Federal Endangered Species Act* (Cambridge, Mass., MIT Press).

ENVIRONMENTAL LITIGATION AND ECONOMIC EFFICIENCY: TWO CASE STUDIES

Charles J. Cicchetti
Robert H. Haveman

The 1960s and 1970s witnessed increased concern in the United States about resource use and environmental quality. In hopes of constraining private-sector activities, citizens turned first to government, and the subsequent legislative response at both the national and state levels was substantial. Water and air pollution legislation, which was passed in several waves, placed increasingly stringent regulations on private sources of effluent. Energy conservation schemes and energy utility tariff reforms sought a reduction in aggregate energy use, as well as the substitution of fuels with larger stocks for those in short supply and a lower level of waste and pollution emissions associated with any energy use. Finally, at the state level, regulatory commissions sought to bring energy use and investment patterns into line with the new conservation and environmental concerns.

The rationale for governmental involvement in these environmental and conservation issues is an economic one. Private-sector decisions are typically based on market prices and costs. Because these values may deviate from their social equivalents as a result of externality or spillover effects, privately used resources may become misallocated, resulting in economic inefficiency. Correcting these market failures requires collective action, and regulations, public subsidies, or charges are thus imposed. This pattern has been true of governmental involvement in the control of air and water pollution. In other areas, what could be worthwhile investments in resources will not be undertaken at all by the private sector because of public goods problems, which explains the government's public works activities in such areas as flood control and navigation investments. In still other areas in which governments have become involved, prices have deviated consistently from marginal costs. For example, under the public utility regulation of so-called natural monopolies, prices have generally been based on average historic costs that can only coincidentally be expected to equal marginal costs.

Often, however, the nature of governmental involvement has neither fulfilled the expectations of citizens nor matched the rhetoric that surrounded the legislation. Governmental enforcement of air and water pollution regulations has been uneven and often ineffective. Legislation typically established an equal allocation but not the most efficient (i.e., least-cost) pattern of pollution reduction. Consequently, uneconomic regulations have been imposed when more efficient alternatives were available, and violations of regulations often went unnoticed or, if they were noticed, unpunished. For example, although there were numerous public investments in flood control and navigation facilities, in many of the projects benefits failed to exceed costs. Indeed, the process used to measure costs and benefits typically ignored a project's adverse environmental impacts and exaggerated its benefits. Many observers concluded that private beneficiaries were being served rather than the public interest. (For a more extensive discussion, see Haveman [1973].) In the utility regulation area, the state commissions continued to defend traditional, average-historical-cost rate making and nearly automatic *ex post* investment approval procedures in the face of evidence that environmental externalities and conservation values were being ignored. Generally, public policies were based on average (fairness) versus marginal (efficiency) principles, and typically the practice failed to meet even its stated objective.

Through public hearings and published critiques, these shortcomings and criticisms of governmental performance were documented, and suggestions for revised and more economically efficient procedures were proposed. By and large, the public-sector response to these was not forthcoming, and those urging more effective measures became frustrated with what they viewed to be the inflexibility and intransigence of government. Increasingly, bodies such as environmental groups, public interest groups, and citizens' lobbies turned to the courts to redress the alleged wrongs. In the case of environmental problems related to energy utilities, these same groups turned to quasi courts in the form of state utility commissions. They were able to use these avenues because unwittingly or otherwise, a variety of legislative acts and legal interpretations of citizens' rights in challenging administrative actions provided for the use of litigation to enforce governmental attention to the full set of social costs and social benefits involved in its decisions.

The use of these legal or quasi-legal processes to ensure rational public decisions, however, has not been a panacea. In bringing litigation against the public sector, those claiming injury often chose a framework of economic efficiency for at least a portion of their case. The government, they argued, had ignored important social values—benefits and costs—in framing its decisions, and relief required that the decisions be altered to reflect these values. Given this framework, the involvement of economic experts in litigation became standard, and both sides in a dispute were forced into complex arguments and estimates of the values that were asserted to be neglected or mismeasured. The courts or other bodies hearing these cases were left to arbitrate and resolve the conflicts.

This paper discusses the effectiveness of relying on legal procedures for the enforcement of sound environmental and economic policy. It begins with a discussion of rationality as the basis for public policy decisions, the role of legal bodies in enforcing these decisions, and the general problems these bodies seem to have with complex and technical analyses of the economic issues involved. Two case studies are then presented to illustrate how issues of economic analysis came to the fore in each case and how courts and regulatory bodies chose to handle the complex issues involved.

It should be noted that our conclusion is not an optimistic one. For a variety of reasons, our examination suggests that the capacity of the courts to provide an effective review of cases involving complex scientific and economic analysis is inadequate. In the first place, some courts seem to be more reluctant to find the government itself negligent or out of compliance with a general rule than to issue such a ruling against other parties to a dispute. In the second place, the material that the courts are required to review is often far beyond the training of court officials or the expertise to which the court has access.

Regulatory institutions, on the other hand, seem somewhat better able to handle cases involving complex issues—for two primary reasons. First, regulatory institutions generally do not consider themselves to be part of the policy-formulating arm of government; rather, they interpret rules that are promulgated by the legislative branch, enforced by the executive branch, and interpreted by the judicial branch. Therefore, regulatory institutions are rarely in the position, as the courts often are, of ruling against the government in which, constitutionally, they are a coequal partner. Second, and perhaps more importantly, the staff of regulatory institutions are usually more highly trained than court staff in the particular issues before them.

The introduction of quantitative economic analysis—benefit–cost studies and the measurement of marginal costs—into the legal, regulatory, and enforcement process is the logical extension of a trend that began in the 1950s. During that decade increasing numbers of public finance and applied welfare economists began developing the theoretical underpinnings and empirical techniques for bringing quantitative estimates of the economic effects of public decisions to bear in the public decision-making process.

Natural resource development (especially water resources development) and environmental policy issues were among the earliest to be subject to the economist's estimates of who was gaining and who was losing because of proposed and actual policies, and the magnitude of their gains and losses.[1] Resources for the Future (RFF) and especially John V. Krutilla and Allen V. Kneese of the research staff were the focal point of much of this activity in both the academic and the policy spheres. (Both of the authors came to RFF during the 1960s to work on the applied analysis of policy measures in the natural resources and environmental areas. Our perspective owes much to this experience.)

The quantitative economic analysis of proposed projects took benefit–cost ratios of less than one to indicate that, whatever else their merits, the projects were diverting resources from high-valued to low-valued uses—that is, net losses of efficiency and economic well-being were being incurred. In addition, estimates of marginal costs that diverged from the cost figures serving as the basis of regulatory choices implied that gains in social efficiency could be achieved by changes in the nature of regulatory decisions.

The initial focus of these analytical studies was the executive branch of government. Agency planners and decision makers, it was believed, would welcome quantitative estimates of the benefits and costs of their decisions and would alter their decisions in response to this added information (see, for example, Hufschmidt and coauthors [1961]). In the Executive Order signed in 1965 establishing the federal Planning, Programming, and Budgeting System (PPBS), this belief was incorporated into planning practice. The order required all agencies to subject their major decisions to the analysis of costs and gains. But the world was more complex than the limited-efficiency perspective of the economists and policy analysts, and little evidence could be found that these analytical studies were actually changing executive decisions. Rather, political considerations seemed to override the guidance of the economists' calculations.

The legislative branch was the next target of the analysts. After holding extensive hearings in the late 1960s and early 1970s,[2] Congress created the Congressional Budget Office and mandated the establishment of an analytical capacity in the General Accounting Office in the Budget Control Act of 1974. In addition, the legal services and environmental movements of the 1970s took the economic analysis of the effect of public decisions to the courts and regulatory agencies when both executive and legislative decisions were found to conflict with the evaluation of social costs and gains.

Achieving Rational Public Decisions

Some Principles for Collective Rationality

Decisions can be considered rational if they are guided by a clear objective, if they seek to attain the objective at the least cost, and if they are made in the presence of accurate information on their effects. This definition applies equally to decisions made by private consumers, firms, and governments. A good case can be made that individual private-sector decision makers have incentives to make rational decisions; indeed, the premise of the entire body of neoclassical economics is that firms and households know what they want, that they choose to secure the attainment of their objectives at the least cost, and that they can obtain information on how various choices affect their objectives. However, no such case can be made for collective decisions. Political considerations, coalitions, checks and balances, power, impossibility theorems, and multiple objectives are all concepts that surround discussions of "public choice," and only under the most constrained conditions can public decisions be naturally viewed as "rational."

The problems of controlling public decisions and ensuring their rationality are of long-standing concern to the two institutions responsible for the oversight of the public sector: the legislatures and the courts. Over the course of time, these two institutions have identified three key elements in the making of a so-called rational decision: (1) a governing rule by which proposed actions are judged, (2) specific facts regarding a proposed action that demonstrate whether the requirements of the governing rule have been met, and (3) an independent review institution that can impose sanctions or enforce compliance when the proposed action is asserted to be inconsistent with the general rule. Let us now consider the first two elements. (In the two case studies that appear later in this paper—one involving a court decision associated with a U.S. Army Corps of Engineering navigation project and the second a state regulatory commission rate-setting decision—we will review the third element.)

Any action or decision made by a government may be measured by its compliance with a rule or principle. For instance, when a law is enacted, it may be measured against the constitutional requirement that government decision making should satisfy two criteria: (1) the goal being pursued by the government should have been made legitimate by the approval of society and (2) the means used to achieve the goal must be "rationally related" to it.[3] This constitutional mandate has been articulated in many ways by the legislature and the judiciary. The courts have long declared that government action that is not related to a legitimate government objective is without "rational basis" and thus "arbitrary" and "capricious." Congress embodied this concept in the Administrative Procedures Act of 1946 and directed the courts to overturn any action that was "arbitrary and capricious." Moreover, the legislature cannot pass a law imposing undue restraints on a specific person in a specific controversy because such a law would represent a discriminatory act that is prohibited by the Constitution.

The executive branch (at both the federal and state levels) plays a double role in the system. As the delegate of the legislature, the executive branch has extensive powers to promulgate general rules. These rules, like the laws on which they are based, are subject to the same constitutional requirement discussed in the preceding paragraph: the rules should be in pursuit of an appropriate goal, and the means selected to achieve the goal should have a rational connection to it. In addition, however, the executive branch acts as a decision maker in specific factual situations, such as the denial or permission of a request by a drug manufacturer to sell a particular drug. In the adjudication of these individual cases, the courts constantly require that the agency apply general rules lest the decision produce a discriminatory result.

Facts become important at all levels of rational analysis. They are important in determining whether a legislative rule has a legitimate purpose and whether the rule's restriction or burden is rationally related to achieving the goal. Once a rule passes the test of rationality, facts again become important in determining whether a specific action meets the requirements of the general rule.

Examples of the general rule and the significance of compliance or non-compliance abound in the food, drug, medical, and consumer products areas. For instance, a general rule has been promulgated that states the following: Products that will injure or kill people cannot be sold. Because virtually any product in some circumstance can cause injury, manufacturers require precision in the general rule before they produce and sell a product. As a result, major efforts are made to define specific rules in the health, safety, and consumer product areas.

Once the general rule has been established—such as a rule that a product cannot be marketed if more than one cancer appears in 10,000 test animals exposed to it, or a rule that no waste dumps should be established within 500 feet of a potable water supply—the question again must be asked: Do the facts involving a particular action demonstrate that the requirements of the general rule have been met? To answer this question requires an evaluation of applicable evidence and a comparison of the evidence with the requirements of the general rule. An affirmative answer indicates that the action is a rational one.

Enforcing Rationality: The Legal–Regulatory System's Role

Who decides whether a rule is rational and whether the facts of an individual proposal meet the requirements of the rule? In the U.S. system of government, these decisions are normally the responsibility of various agencies of government. But no agency can be totally consistent and honest in judging whether its actions are in compliance with particular rules. Under rational decision making there must be an independent institution to decide when a rule has been violated and to enforce compliance. This function has been assigned to the courts and also to administrative agencies that function as judicial-like bodies (e.g., regulatory commissions).

Because the government relies heavily on voluntary compliance with general rules, the judiciary's role is primarily passive. In the absence of claims by third parties that proposed agency actions violate the rules (i.e., either legislative acts or the Constitution), the judiciary is not involved, and governmental decision making proceeds smoothly. An important reason why voluntary compliance occurs, however, is the public's knowledge that, if required, the judiciary will enforce compliance with the rules. Thus, factual disputes about whether an empirical basis for a rule exists, challenges that the rule is "arbitrary and capricious" (and hence, irrational), and disputes over whether an action violates the rule are to be resolved in court.

Courts and other judicial bodies have devised certain procedural rules for resolving such disputes, rules that embody society's collective sense of fair play and that are intended to ensure a high degree of factual accuracy in the ultimate decision. These procedures, whose purpose is to ensure that all concerned will accept the decision of the judicial body, are intended to satisfy three basic requirements:

1. The facts and reasoning relied on by both proponents and opponents of an action must be revealed. A factual claim is typically the end product of a series of logical deductions from certain facts or data. Each side in a dispute should be able to replicate the data and the chain of logic that lead to the other's factual claims.
2. The factual or analytical defects in one side's evidentiary claims, defects that may support the other's position, should be known. It is a basic procedural principle that each party in a dispute has a duty to disclose all evidence relevant to the ultimate factual issues—including evidence that may affect his or her own position adversely.
3. As the final arbiter, the judicial body must be able to understand the evidence before it and be able to determine the validity of a factual conclusion—if necessary, by replicating the data and the logic leading to the conclusion.

If the judiciary is to perform as an independent institution making substantive policy decisions and enforcing the use of rational decision making on the legislative and executive branches, two requirements must be met. First, the judicial body must have the necessary technical capability and intellectual capacity and energy to understand the issues that are being litigated and to reason through the data and logic supporting the factual claims. Second, the body must be able to replicate the data and logic underlying a factual conclusion. This requirement places the burden squarely on the proponents of a position to provide a clear, step-by-step presentation of the data and logic supporting that position and the links between them.

These procedural ground rules of fairness, which are often more formally attired in due process and statutory procedural language and labels, are essential if the judiciary is to have credibility in resolving factual disputes about whether an action complies with or violates a general rule. The failure to adhere to these and similar procedural ground rules invites irrational decision making and threatens to destroy the nation's confidence in the institutions of the judicial system.

Enforcing Rational Choices by Government

The judicial system must apply these principles in both private litigation and in proceedings involving the government as either plaintiff or defendant, as well as in small and simple disputes and major, complex ones. Yet it is in those proceedings in which (1) the issues are technical and complex and (2) the subject of litigation is an actual or proposed action by government that judicial bodies, especially courts, have the most difficulty living up to their own standards. This is so despite the fact that the history of Anglo-American legal development is largely the creation of general rules designed to constrain irrational, ad hoc action by government.

Several factors have impeded the role of the judiciary in ensuring that the actions of governments conform to general rules. We will mention only two here. First, by its very position as a rule maker and enforcer, the government seems to have acquired special standing in legal disputes. In some parts of the judiciary, there is more reluctance in ruling that the government is negligent or out of compliance with a general rule than in issuing such a ruling against other parties to a dispute. In these cases, the judicial body tends to confuse two constitutional precepts: one, that the body should be slow to interfere in legislative and executive branch decision making, especially if it appears that these branches are in agreement; the other, that the body should vigorously enforce the law. Although courts, in particular, have a mandatory duty to enforce the law, even if that requires enjoining governmental action, the very institutional position of government appears to have impeded this function.

Second, in addition to the judiciary's duty to review the factual basis of a proposed action and to test its consistency with the general rule, it is essential that a judicial body have the capability to sort through and assess the evidence laid before it. This function is sorely impeded if the material, by design or otherwise, is of a complexity or technical character that is far beyond the training or expertise to which the judicial principals have access. In fact, in increasing numbers of instances, courts especially have become constrained by what we may colloquially describe as "judicial math block." Judicial principals cannot rationally undertake a responsible and thorough review if they are unable to understand the factual evidence presented by the proponent or opponent of a proposed action.

The consequences of judicial math block are staggering. In the sphere of private disputes—such as patent and antitrust litigation—opposing parties are often faced with the realization that the presumably independent weigher and sifter of facts has little understanding of the evidence being presented by either side. Moreover, this comprehension gap tends to widen when expert testimony, as it often does, obfuscates the facts instead of educating the court. In disputes over proposed government actions, the situation can be even worse. In these cases the government often produces thick volumes of government-sponsored research performed by prominent consulting firms. In part because of the government's unique position, the judiciary may refuse to probe even elementary calculations, finding it easier and of less potential embarrassment to simply accept the government's case and, therefore, its claim.

Because of these two problems—confusion over the judiciary's constitutional duty to review factual compliance with general rules and the so-called judicial math block—the resolution of disputes involving complex scientific and economic issues to which the government is itself a party can, in parts of the judicial system, be characterized as approaching a state of scientific anarchy. Moreover, government lawyers and experts, along with those private interests that benefit from the governmental decisions, have an incentive to exacerbate these problems. Hence, the government is increasingly found to cloak relatively simple calculations and logic in complex formulations and then bury them in voluminous reports. Even if the data and logic chain can be simply understood, the government seeks to give the calculation an aura of scientific complexity. Resorting again to the colloquial, the government's case is all too often based on its belief that the court can be "snowed." As a result, the probing skepticism that is so critical to a rational decision process becomes imperiled.

Although courts more readily tend to fit this characterization of a lack of effective scrutiny of governmental decisions and the basis for them, parts of the judicial–regulatory system also appear to apply similarly deferential attitudes to private interests whose activities and decisions are to be scrutinized. Again, the same pattern of obfuscation and the use of excessively technical presentations exacerbate this problem. Our case studies will examine two judicial-type bodies and their handling of technical issues. Staff capability to interpret technical material and complex formulations distinguishes the two bodies—a court and a regulatory commission.

Case Study 1: Environmental Litigation Over Public Navigation Investments

Statutory and Historical Background

Virtually all of the proposed additions to or improvements in the inland navigation system of the United States are the responsibility of the Army Corps of Engineers. Since the passage of the Flood Control Act of 1936, the Corps of Engineers can recommend a project only if it complies with the general rule that the economic benefits exceed the economic costs, which explains the heavy reliance on benefit–cost studies in this area. Important laws and regulations pertaining to the conduct of the Corps' benefit–cost studies include the U.S. Department of Transportation Act of 1966, the National Environmental Policy Act (NEPA) of 1969, the River and Harbors and Flood Control

Act of 1970, and the principles and standards promulgated in 1973 by the U.S. Water Resources Council.

In any litigation over the environmental effects of federally sponsored navigation projects, a key question is whether the analysis accompanying a proposed project fulfills the analytical requirements of NEPA. Although the focus of such analysis is on their environmental effects, a standard benefit–cost analysis must be included in the evaluation because NEPA requires a thorough comparison of alternatives to the proposed action, including the alternative of doing nothing. Consequently, every evaluation of a navigation project requires substantial data manipulation and statistical analysis. In particular, the Corps of Engineers' regulations, in accordance with statutory requirements, prescribe that the following set of specific calculations be developed in connection with estimating the benefits and costs of public investments in navigation facilities:

- Potential traffic is identified by means of a field survey and an analysis of existing freight movements in the tributary area.
- These potential movements are screened to eliminate those that are not susceptible to diversion to the proposed waterway.
- For those movements that remain, the actual rate that shippers are paying for transportation is compared with the expected rates (all other associated costs being considered) on the proposed waterway.
- The estimated per ton savings to shippers for each movement is multiplied by the base-year tonnage to derive base-year benefits.
- The tonnage and estimated savings to shippers for each movement are projected over the life of the project (normally fifty years).
- The present value of project costs and benefits is derived by discounting, using the long-term borrowing rate of the federal government.

These procedures are explored in detail in Haveman (1972). One of the disconcerting aspects of these analysis requirements is that the benefits measure savings in transportation charges to waterway shippers and not benefits to the nation. To be sure, the benefits to shippers might be a close approximation of the benefits to the nation if shippers' costs in the various competing media— the railroads, the highways, and the waterways—were being determined in an open competitive market; in that case, the rates paid by shippers would approximate the nation's marginal costs of providing the services. But during the 1970s, the rates were being fixed by regulatory bodies pursuant to standards that in many cases were far removed from marginal costs. Thus, the fundamental basis of the comparison prescribed by the statute was questionable in itself, although this issue was not raised in the case reviewed in this section.

Another large question raised by the statute's requirements had to do with the familiar problem that plagues most benefit–cost studies, namely, the appropriate discount rate for use in calculating present values. The long-term borrowing rate of the federal government is clearly inappropriate for discounting purposes. That rate might or might not represent the opportunity cost of capital to the government, but it fails totally to reflect the risks to the government that are inherent in estimates of future cost–benefit streams.

To return to our case study, prior to the passage of NEPA in 1969, the Association of American Railroads had challenged a number of specific waterway projects. Directing its efforts toward securing legislative relief, the association sponsored studies that were essentially economic re-analyses employing different data and assumptions and quite naturally arriving at conclusions different from those of the Corps. It was only after NEPA's passage, however, that legal action was taken to actually stop navigation projects. Early suits were initiated by a private environmental organization known as the Environmental Defense Fund against the Cross Florida Barge Canal, the Tennessee–Tombigbee Waterway, and the Tennessee Valley Authority's Tellico project. (The Tellico, or "snail darter," case is discussed by Davis [in this volume].) Because of their limited resources, the plaintiffs modeled their challenges on the earlier efforts of the Association of American Railroads; the association's assumptions were altered and its discount rates modified, but the basic data on potential traffic and rate differences were left unchallenged by the environmentalists.

The Locks and Dam No. 26 Case

The first major effort to subject a Corps of Engineers economic study of a navigation project to detailed scrutiny began in August 1974 when the Sierra Club, the Izaak Walton League, and the Western Railroad Association (a group of twenty-one railroads) brought suit in a federal district court to stop the proposed replacement of the structure known as Locks and Dam No. 26 in the Upper Mississippi River system. In this case, the plaintiffs challenged both the basic data and the statistical procedures used by the Corps in its benefit–cost analysis.

Locks and Dam No. 26 is located on the Mississippi River north of St. Louis, Missouri, below the confluence of the Illinois and Upper Mississippi Rivers. All Illinois and Upper Mississippi River barge traffic originating or terminating at or below St. Louis must pass through the structure, making it a key facility on the waterway system of the Mississippi and its tributaries.

In 1969 the Secretary of the Army approved the expenditure of $203 million to replace the existing Locks 26 structure with a new dam and two locks to be located two miles downstream of the present site. The Corps had justified the replacement proposal in a report issued the previous year (U.S. Army Corps of Engineers, 1968); it subsequently buttressed that justification with four other major reports (U.S. Army Corps of Engineers, 1972, 1975, 1976a, 1976b), all purporting to demonstrate that replacement of the existing Locks 26 was the best alternative.

The series of feasibility reports on the project are noteworthy for their widely different methodological, statistical, and evaluative approaches. The average annual costs charged to the replacement project over a fifty-year life span vary from $9.5 million to $52.3 million, and the benefit–cost ratios reported in the various studies range from 1.5 to 8.6 (Carroll and Rao, 1978). These differences can be attributed to a number of factors. For one thing, some of the reports covered only the direct effects of increasing the capacity of Locks 26, while others introduced effects that included other investments as well. In addition, the reports used different estimates of existing and proposed lock capacity, different forecasts of commodity traffic, different estimates of present and future rates, and different rules for allocating costs in the fixing of rates.

In the lawsuit filed in August 1974, the plaintiffs made a number of claims: (1) that the project had not been authorized by Congress; (2) that it was part of a much larger plan of the Corps of Engineers for expanding the capacity of the waterway system above Locks 26; (3) that the Corps had failed to consider alternatives to the proposed plan that might have had more favorable effects; and (4) that background data had been withheld from the plaintiffs, so the benefit and cost computations could not be verified.

In September 1974 the U.S. District Court in the District of Columbia upheld the plaintiffs on virtually every allegation and enjoined the Corps from proceeding with the project. In 1975 the Corps prepared another report that, along with various supplements, provided the basis for the economic component of the testimony introduced by the plaintiffs in a trial held in September 1979.

The purpose of the Corps' benefit–cost study was to determine whether the benefits from the new facility (that is, the savings to shippers in transport charges) exceeded its costs; both benefits and costs were stated in terms of present value. To estimate the savings in transport charges, it was necessary to identify the traffic movements that were expected through the facility in each future year. Then, for each movement, the transportation mode that would have been used if the proposed facility did not exist had to be identified. Finally, the difference between the two alternatives in freight charges and other associated costs had to be estimated; and this difference—representing so-called savings—had to be aggregated for all movements in order to yield an estimate of the total benefits to shippers. These calculations had to be made for each year of the fifty years for which the proposed facility was expected to be in existence.

The analyses entailed quite complex calculations. In effect, a demand curve had to be constructed both for the existing facility and for the proposed new facility expressing the relationship between savings per ton and the use of the facility. But savings per ton was itself a complex calculation: it would be affected by the relevant rate structures as well as by the level of use of the facility because of the effects of such use in delaying shipments and generating other costly incremental charges. These marginal demand and cost functions had to be solved simultaneously to obtain the level of facility use in each year. Because both the demand and the cost function estimates for any given year depend in turn on estimates of a large number of underlying variables,[4] these estimates offered room for substantial differences in professional judgment.

Obviously, a key element in the calculation of both the demand function and the cost function is the charges borne by the shipper. To obtain an accurate description of shipper costs with and without construction of the project, it was necessary to know the actual origin and destination of the movements as well as the barge rates. For many movements, the Corps never revealed the actual origins or destinations, making it impossible either to develop independent calculations of shipper costs or to verify the Corps' calculations. Furthermore, some barge rates were developed under secret contracts with private consulting firms; they were generally unpublished and thus could not be verified.

To assist in the numerous arithmetic calculations required to determine project benefits, the Corps developed a "simulation" model; however, the step-by-step logic and arithmetic computations behind each data point were not disclosed in any public document. It required massive discovery motions and extended negotiations by the plaintiffs' lawyers before the Corps supplied sufficient information to allow the plaintiffs' expert witnesses to understand that much of what the Corps was doing was simple arithmetic. And even after this partial disclosure, the Corps' withholding of key information on origins, destinations, and barge rates effectively foreclosed the possibility of critical review. Thus, the plaintiffs were faced with a situation in which the government agency made a bald claim—that the project would reap $86 million in annual benefits—with virtually no explanation of the data and logic underlying its claim.

The Corps also pursued another practice that tended to impede any independent appraisal of its claim: the bare-bones description of its calculations was buried in literally thousands of pages of essentially irrelevant text. And the actual calculations were nowhere in the official reports. Yet when the question of the empirical basis for the $86-million annual benefit claim came before the courts, the Corps pointed to the several-foot-high stack of irrelevant text as substantiating its benefit claims. Finally, pervading the entire Corps strategy was the tendency to make simple calculations appear complex and mathematical.

The opponents of the project thus went to the courthouse with several strikes against them. Their hope was that the trial judge would surprise the Corps and demand the specific data behind the $86-million claim. The essence of the position of the Locks 26 opponents was that professional economists and transportation experts retained by the plaintiffs were unable to replicate or verify the data leading to the Corps' benefit claim of $86 million. Moreover, in the absence of disclosure of the data to the plaintiffs and their experts, the court had no rational basis for accepting or rejecting the $86-million benefit claim and thus could not perform its function of determining whether the facts of the Corps' proposal met the requirements of the general rule that benefits should exceed costs. Based on the limited disclosure made by the Corps, the plaintiffs also went beyond these basic positions and presented a number of arguments demonstrating that key empirical assumptions made by the Corps were either inconsistent with reality or based on insufficient data to merit acceptance by any decision maker.

The Locks 26 trial held in federal district court in Washington mirrored the problems that were identified earlier in this paper involving judicial review of factual disputes that relate to scientific or economic issues. In a trial with complex direct and cross examination of economic, environmental, and engineering testimony involving hundreds of millions of dollars of potential public and private losses and costs, the judge assigned five days to the entire trial. In addition, the trial judge did not ask for, nor did he receive, an explanation of the $86-million annual benefit claim—even though the Corps' opponents repeatedly requested this material. After the trial, the judge refused to decide critical questions of fact, instead deferring to the "expertise" of the Corps of Engineers.

Finally, after failing to consider the statistical models provided by the plaintiffs as well as the plaintiffs' critique of the agency's data, the judge ruled in favor of the government. Verification of the accuracy of the statistical data, the appropriateness of models designed to analyze the data, and the facts supporting the assumptions driving the model and the analysis were not accepted as issues on which a legal ruling should be made. Rather, the court considered the numbers, models, and favorable conclusions produced by the agency as sufficient to rule in favor of the government. This conclusion was accompanied by several feet of paper that purported to document the facts and logic supporting the Corps' proposal but that actually contained primarily irrelevant filler. In the end, the phenomena discussed above—judicial math block, the failure of the court to compel compliance to the general rule, and an unwillingness to challenge government "expertise"—played the crucial role in judicial resolution of the issue. Admittedly, a single case study is not sufficient to establish general conclusions; yet other environmental cases of this type suggest the accuracy of such a generalization.

This particular litigation shows more about the difficulty of introducing economic analysis and data into the judicial process than it does about the success of the work of analysts on the ultimate policy outcome. Such a melancholy picture, however, focuses only on the direct outcome of the litigation. To be sure, the dam and the locks were ultimately authorized and constructed. Yet in the process of the congressional debate that followed the litigation, legislation was passed that imposed a user charge in the form of a fuel tax on the barge operators who heretofore had received the benefits of the water right-of-way free of charge. Future navigation projects and their maintenance will be financed in part by revenues from this tax. This longer-term and indirect effect of the debate stimulated by the litigation—and its focus on economic costs and gains—is likely to have a far greater effect in constraining the allocation of resources to wasteful activities than that represented by the decision on Locks and Dam No. 26. Indeed, should such a cost-sharing requirement be widely adopted, future demands by private interests for taxpayer subsidies of inefficient facilities that provide benefits to themselves may well be severely blunted.

Case Study 2: Environmental Litigation in Electricity Pricing

Institutional and Historical Background

Each state has a regulatory body that determines the pricing policy, income, and investment behavior (to varying degrees) of the nation's privately owned energy utilities (both electric and natural gas). In the United States, such regulation covers approximately 93 percent[5] of all retail electricity sales and 94 percent[6] of all retail natural gas sales. As a general rule, state regulatory commissions are required to ensure that the public interest is protected; thus, they serve as quasi-judicial bodies. Economists interpret the need to protect the public interest as meaning that state regulators should adopt economic efficiency as their objective or goal.

Privately owned energy utilities typically have been granted a franchise that ensures monopoly status in their specifically chartered service territory. The premise for this form of economic governance has been the historic fact or belief that energy utilities were "natural monopolies." In other words, because average total costs were expected to decline as output expanded, a single supplier was expected to produce electricity and natural gas services at lower unit costs than a competitive industry. In a competitive environment, firms would, by necessity, either invest in duplicate facilities or fail to achieve the available economies of scale, or both.

Generally, state legislation gives regulators three tasks or general rules to follow:

1. The earnings, or realized return on investments, should be restricted to levels comparable to returns in competitive industries.
2. Utilities have a duty to serve all customers who are able to pay for the service; in other words, access to utility service cannot be denied.
3. The prices charged for utility services must be cost-based and nondiscriminatory.

To accomplish these general rules, state regulatory authorities have evolved an institution known as a rate case proceeding in which there are four distinct steps. These steps can be understood by considering the following algebraic equation, which provides the central policy focus of rate regulation:

$$\text{\begin{array}{c}Revenue\\requirements\end{array}} = \text{\begin{array}{c}Operating\\costs\end{array}} + \left(\begin{array}{ccc}\text{Total cost of} & & \text{Authorized}\\\text{rate base less} & \times & \text{rate of}\\\text{depreciation} & & \text{return}\end{array}\right)$$

With this equation as a focal point, the first two steps in a rate case proceeding can be described:

1. *Cost-of-service studies:* The utility combines mathematically complex accounting, engineering, and economic data and studies with generally accepted rules and precedents to determine how much it expects to sell (or, in some states, did sell) in a "test" year. This sales estimate is one of the chief factors in determining the operating cost component of a utility's regulated revenue requirement. The second component of a cost-of-service study is to determine the original cost of past investments and then subtract what has been depreciated in value (or the rate base) for rate case purposes. Because this step, especially today, may involve a review of the prudence or reasonableness of past as well as ongoing investments, this apparently straightforward step is technically and politically complex.

2. *Rate of return:* Public utility regulation also determines the appropriate rate of return on the rate base. Regulators calculate a weighted-average rate of return whose two principal components are (1) the various interest rates to be paid on each outstanding, historically issued debt instrument, and (2) the allowed return on equity. Typically, the latter is determined by regulators after a technically complex and fully litigated proceeding. The standards used to establish the often politically charged rate of return on equity include complicated financial and statistical analyses as well as economic comparisons of comparable earnings. In recent years regulators have sometimes adjusted the debt and equity weights used to determine the weighted-average rate of return if, after an evidentiary proceeding, they conclude that the "actual" capital structure is inconsistent with the public interest. To reach such a conclusion obviously involves the review of a significant amount of contradictory expert evidence.

The last two steps in a rate case proceeding involve the translation of the results of the previously defined revenue requirements formula into prices. This translation requires that the sources for revenue be assigned their respective roles and that a set of prices be designed. To some extent the distinction here is artificial because step 3 determines the amount of revenue to be realized from each source and the pricing design (step 4) selects the pricing structure to achieve that target. Clearly, these must be joint activities; nonetheless, they are often described as distinct steps—the allocation of costs and rate design.

3. Allocation of costs: This step is more accurately described as spreading revenues. Utility services are not simply sold in the retail market at a standard unit price. Instead, utilities and their regulators have established various categories of service. Typically, each category either has some common user characteristics (e.g., all commercial customers may be lumped together) or some similar cost characteristics (e.g., all residential space-heating users). At last count, there were about thirty different varieties of cost allocation formulas that have been proposed to allocate costs to the various customer categories in U.S. rate cases. The debate on which method to adopt and how it should be applied in a specific rate case is often very complex.

4. Rate design: Until recently, the principal feature embodied in electric and natural gas tariffs was a form of volume discounting referred to as declining block rates. Three rationales have been suggested for this pricing practice: (1) with decreasing costs, promoting use should lower unit prices; (2) collecting revenue up front in the first block was either based on cost (i.e., it was thought that lower-volume users cost more to serve) or helped to ensure the recovery of the authorized revenue requirement; and (3) according to the economic theory of a natural monopoly, in order for firms to price marginal use at marginal costs and not go out of business, some device must be found to generate additional revenue without unnecessarily influencing marginal use.

The Madison Gas & Electric Case

The first major effort to apply economic efficiency arguments to an electricity rate design case began in 1972 when the Environmental Defense Fund (EDF), a national group, joined forces with two local groups—Wisconsin Environmental Decade and Capitol Community Citizens—in an attempt to require Madison Gas & Electric Company (MG&E) to replace declining block pricing with marginal cost pricing. The MG&E case turned out to be a landmark state regulatory decision. One reason for this prominence was that all of the major electric utilities in Wisconsin decided to join forces to oppose the intervenors, who in this case were associated with environmental advocacy groups. Additionally, Wisconsin has been recognized as a policy trend setter among state utility regulatory commissions.

The utilities' first set of arguments had three components. First, tradition required utility commissions to measure actual historical costs. Second, marginal cost was an interesting, perhaps irrelevant, academic concept (later, the argument against marginal costs was based on the concept of "second best" problems). Third, customers' demands were not sensitive to price changes. Indeed, the utilities' initial argument maintained that demand was completely price inelastic: in such circumstances, regulators should simply ensure that fair and accurate costs were allocated to each customer's bill and not be bothered by tariff structures.

The environmentalists' intervention involved a detailed set of issues. Their arguments were complex and involved both the concept of marginal cost and its measurement in the electric utility case, in which environmental costs were associated with the production of the electricity output. The intervenors attempted to communicate these concepts and procedures in simple, straightforward terms. In considering the question of estimating the marginal costs of services, the utilities argued that marginal cost was a theoretical concept and difficult to define in terms that allowed for "practical" measurement. The intervenors, by contrast, contended that efforts to move the process closer to achieving economic efficiency (i.e., maximizing net social welfare) were more important than the ease of measurement of average accounting costs.[7]

The second component of the environmental intervenors' case concerned their contention that prices mattered and that demand would respond to price changes. Consequently, investments in capacity and the magnitude of environmental costs, both of which are associated with generating electricity, can be expected to be influenced by pricing decisions. This second point directly follows from the first. Economic efficiency requires that prices signal the full marginal cost, that is, the real resource cost, of electricity. Yet this premise has not been the basis for regulation in the past. In the view of intervenors, precedent is not required to correct past mistakes. They contended that the position of the utilities represented a "nothing should ever be done for the first time" perspective.

With these two straightforward arguments, the intervenors attempted to demonstrate the technical feasibility of their underlying arguments—that marginal costs could be measured and that price responsiveness did exist. They introduced empirical studies into evidence, developed and expanded simple procedures for estimating marginal costs, and presented computer simulations to indicate the substantial difference that the cost concept and measurement would make. As the commission staff and the members of the commission began to respond to these principles and evidence, the Wisconsin utilities replaced their original consultants and organized their own team of economic experts. Their position seemed to be that if economics and econometrics are to replace traditional accounting approaches for defining prices (i.e., rate design), then the utilities and their experts would show how it could be done properly.

The economists on both sides of this regulatory hearing agreed readily on three issues:[8]

1. Because of the significant time-of-use and voltage differences associated with delivering electricity to the meter, inverted block pricing made no more sense (in terms of economic efficiency) than declining block pricing.

2. The price elasticity of demand was not zero and may well involve both types of customer and time-of-use components. Thus, it would be desirable to examine both the effects of time-of-use prices (especially because costs were expected to vary rather dramatically with time of day, week, and season of use) and the price responsiveness of different customers on tariff design.

3. Any external social costs should be included in the estimates of the utilities' marginal costs.

But despite the economists' agreement, two tasks remained. First it was necessary to implement a practical marginal cost pricing system, which meant that it was necessary to estimate marginal costs. Both the United Kingdom and France had attempted to develop and implement marginal cost pricing arrangements in their electricity tariff and marketing arrangements, but the written research was specific enough to each country's institutions to make the transfer of anything but general approaches and potential pitfalls subject to doubt. With the support of the National Science Foundation, a cross-national conference of economists and engineers from these nations was organized to facilitate direct communication among experts. As a result of these discussions, much analytical and institutional information relevant to the initiation of marginal-cost–based electricity tariffs was obtained. Most fundamentally, however, these discussions revealed that a close surrogate of true marginal costs was already being calculated and used by the Wisconsin utilities in their system dispatching and planning functions. Utility engineers had been estimating short- and long-run marginal costs, without actually calling them that, in order to respond to their charge to minimize short-run operating (dispatch) costs and long-run system expansion costs. Because in Wisconsin the rate-making function was largely the domain of utility accountants charged with collecting authorized revenue requirements, the marginal cost calculations of the engineers remained unknown to the rate makers. Finally, however, the merging of the two disparate information sources that already resided in the Wisconsin utilities made possible the implementation of a marginal cost tariff arrangement.

The second task was to draft a rate order. The Wisconsin regulatory commission, in its Final Order, Docket 2-U-7423, found marginal cost pricing to be the correct conceptual basis for pricing utility services in Wisconsin[9] and required all Wisconsin utilities to estimate marginal costs, prepare tariffs, and develop implementation plans before initiating any request for a rate increase. At the time, time-of-use demand studies were not available for any class of customers, although they were necessary in order to estimate the consequences of full marginal cost pricing (for example, the practical issues associated with deciding the number of different prices to be charged by time of day, week, and year and the degree of responsiveness in them to customer responses).

Equally important was the fact that metering costs were uncertain. The final order, which drew heavily on the theoretical, empirical, and implementation knowledge introduced into the case by the intervenors, deferred these matters to subsequent future rate case filings.

The MG&E rate case, then, is an example of a situation in which the application of economic analysis and data was able to directly influence the direction of policy. There are several reasons for this. First, the increasing relative scarcity of energy fuels and the environmental consequences of using them were high on the research agenda of a large number of economists and engineers at that time; thus, a wealth of talent and expertise involving similar backgrounds and training could be brought to bear on this issue in a relatively short time. Second, the basic questions at issue were relatively straightforward and had been worked on for decades within the economics fraternity. Does the consumption of electricity depend on its price? How should electricity tariffs be designed in the face of decreasing direct costs and increasing environmental costs? How can the utility revenue constraint be reconciled with economically efficient pricing rules? Third, unlike the case of the court system, the regulatory commission is a quasi-professional organization. Its members are often individuals with substantive knowledge in the area, and they have access to economists and engineers who serve as their staff. In such an environment, analytic information perhaps has its best chance of influencing decisions. Finally, unlike the Locks and Dam No. 26 case, the issue at stake in the electrical pricing case involved two private parties with divergent interests and objectives. In the water resources case, the defendant was the federal government, and the court was placed in the position of second-guessing and overriding the executive or legislative branches of government.

The MG&E case had a number of other important side effects as well. In the process of the hearings and the testimony, computer programs for calculating marginal cost that could be brought into a hearing room were developed. In addition, the Public Utility Regulatory Policies Act of 1978 was passed by Congress requiring the collection of information that could be used to calculate marginal cost and time-of-use tariffs. A series of federally financed time-of-use electricity tariff experiments were also undertaken throughout the nation. Perhaps most importantly, the National Association of Regulatory Utility Commissions required the nation's electric utilities to embark on a multiyear study to resolve the various economic and statistical questions related to tariff reform. In short, a large number of very extensive electricity tariff reform research agendas were set in motion by the MG&E case.

Accomplishments in the policy arena based on the issues raised by this case have been no less impressive. They began with various reforms in electricity tariff setting, starting with the virtual elimination of volume discount pricing. Eventually, these reforms encompassed time-of-use pricing and costing, marginal cost pricing for cogeneration, and long-range least-cost planning, including conservation. At present, there is a direct link from the MG&E case with the debates concerning industry restructuring through the introduction of competition in generation and market-sensitive pricing.

Conclusions

All too often, we find that major governmental decisions are being made on the basis of factual premises that are either inadvertently erroneous or deliberately false. But the empirical error in these false factual premises is being obscured by several interrelated phenomena: when litigation is involved, factual issues and proofs underlying governmental decision making are often deliberately made unnecessarily complex so as to confuse and discourage probing analysis—which a court is usually unwilling to permit anyway when the defendant is another branch of the government.

The key constitutional institution designed to curb irrational and arbitrary governmental decision making—the judiciary—falls easy prey to such confusion because of a lack of training in assessing complex empirical issues and an apparent misunderstanding of its constitutional duty to treat the legislative and executive branches like other defendants and, therefore, to check arbitrary action.

Although the law requires that a wide variety of public actions be supported by empirical investigation and statistical analysis, it is ironic that, in a growing number of cases, these requirements have tended to hinder the judicial process rather than assist it. The statistics, computer models, and probabilities in these government-sponsored studies have appeared sufficiently forbidding to the courts that extreme deference has been given to the government's position. The conclusions of government-sponsored reports have been assumed to rest soundly on analysis that has been accepted as accurate, pertinent, and appropriate by other more knowledgeable and more expert government officials. When such a presumption has been challenged by third parties, the government's use of scientific jargon, computer modeling, and seemingly complex statistical analysis has often managed to keep the debate obscure—notwithstanding that the underlying issues have sometimes been relatively simple and straightforward.

Increasingly, then, courts appear to be failing to discharge their constitutional role as neutral institutions to review complex disputes that involve the government's application of federal rules. That failure is due partly to the judiciary's confusion concerning its responsibilities vis-à-vis the other two branches of government and partly to the judiciary reluctance to probe into the statistical analyses presented by government defendants.

Yet our second case study suggested that a regulatory body, in contrast to a court, may operate quite differently. The principal motivation of the environmental intervenors was to propose that pricing decisions must consider the marginal costs of service. These arguments implied that the intervenors' lawyers and economists requested state regulators to evaluate the case based on technical economic evidence. For advocates of utility rate reform, this premise turned out to be largely true, at least in the case we have examined. Although the regulatory process has numerous weaknesses—in particular, a history of showing deference to private interests whose activities are to be regulated— it has made far better use than the courts of expert staff capabilities to sort through complex and technical arguments. If courts are to discharge their constitutional role as a neutral institution in regard to governmental actions, they must be encouraged to find a means of developing the capability— perhaps through court-appointed experts or referees—of understanding, reviewing, and sorting out the "facts" of cases, especially when such facts are complex and technical. Because an understanding of these kinds of facts requires that the computer modeling and statistical analyses of both sides to a dispute be proved, the introduction of these competencies into court processes seems essential.

Finally, as discussed throughout, both case studies demonstrate the potential value for policy formulation of the application of economic analysis to current public issues. They illustrate the most recent stage of evolution of this practice: the introduction of high-quality research and analysis into the adversarial arena of litigation and regulatory enforcement activities. As such, they represent the latest step in the process of bringing economic analysis to bear on policy deliberations, a process whose roots extend—and the responsibility for which can be largely attributed—to the research, insights, and commitment of John Krutilla and his associates at Resources for the Future.

Acknowledgments

This paper builds upon and extends earlier work of the authors, including the following: J. L. Carroll, R. H. Haveman, and J. V. Karaganis, "When Complex Facts Threaten Court Reviews: Litigation over Navigation Projects," *Journal of Policy Analysis and Management* vol. 2, no. 3 (1983) pp. 418–431; and C. Cicchetti, W. Gillen, and P. Smolensky, *The Marginal Cost Pricing of Electricity* (Cambridge, Mass., Ballinger, 1977). We would like to thank V. Kerry Smith and Carol May for their most helpful input and comments.

NOTES

1. The earliest of these applied welfare analyses were in the water resources area. Three books appeared at about the same time (Eckstein, 1958; Krutilla

and Eckstein, 1958; McKean, 1958) and among them laid out a set of principles, an approach, and examples of how to proceed, all of which had a substantial effect on the economics profession.

2. See U.S. Congress, Joint Economic Committee (1969). Krutilla (1969) also contributed to this effort.

3. *Thompson* v. *Gallagher*, 489 F.2d 4434 (5th Cir. 1973).

4. For example, the demand function over time depends on the following: projections of traffic growth over time; projections of traffic composition, by commodity and by origin–destination, over time; projections of barge rates and rates in alternative transportation modes over time; estimates of the rates applied by the competing modes of transportation, broken down by product and origin and destination; and estimates of associated shipper costs when moving commodities by various modes, such as costs for transshipment, inventory, and damages. Calculating the cost function was no less complex. It entailed estimates of the relationship of delay times to tonnage levels; estimates of inventory costs, by commodity; estimates of equipment rental costs such as trucks, barges, and boxcars; and estimates of the patterns of sequencing barge tows and assembling barges and towboats into flotillas.

5. Total retail $=$ investor owned $+$ cooperatives $+$ municipals
   % State reg $=$ (investor owned)/(investor $+$ cooperatives
   $+$ municipals)

6.
$$\text{Total retail} = \frac{(\text{distribution} + \text{integrated} + \text{combination}}{(\text{distribution} + \text{integrated} + \text{combination} + \text{municipal})}$$

7. Ironically, the proceedings revealed that another reason why accounting costs were not precisely measured was because there were nearly thirty conflicting approaches to analyzing embedded average costs.

8. The observation that agreement was relatively easy should be considered in relative terms. The actual decisions involved many months, pages of testimony, and extensive hearings. Nonetheless, *ex post* examination of the record indicates clear agreement among the economists on the principles involved.

9. See Madison Gas and Electric Co., 5 PUR 4th 28 (Wisconsin, 1974).

REFERENCES

Carroll, J. L., R. H. Haveman, and J. V. Karaganis. 1983. "When Complex Facts Threaten Court Reviews: Litigation over Navigation Projects," *Journal of Policy Analysis and Management* vol. 2, no. 3, pp. 418–431.

Carroll, J. L., and S. Rao. 1978. "Economics of Public Investment in Inland Navigation: Unanswered Questions," *Transportation Journal* vol. 17, no. 3, pp. 27–54.

Cicchetti, C., W. Gillen, and P. Smolensky. 1977. *The Marginal Cost Pricing of Electricity* (Cambridge, Mass., Ballinger Publishing Company).

Eckstein, Otto. 1958. *Water Resources Development: The Economics of Project Evaluation* (Cambridge, Mass., Harvard University Press).

Haveman, Robert H. 1972. *The Economic Performance of Public Investments: An Ex Post Evaluation of Water Resources Investments* (Baltimore, Johns Hopkins Press for Resources for the Future).

————. 1973. "Efficiency and Equity in Natural Resource and Environmental Policy," *American Journal of Agricultural Economics* (December) pp. 868–878.

Hufschmidt, Maynard, Julius Margolis, and John Krutilla with Stephen Marglin. 1961. *Standards and Criteria for Formulating and Evaluating Federal Water Resource Development*, in Report of Panel of Consultants to the Bureau of the Budget (Washington, D.C., June 30).

Krutilla, John V. 1969. "Efficiency Goals, Market Failure, and the Substitution of Public for Private Action." Pp. 277–290 in U.S. Congress, Joint Economic Committee, *The Analysis and Evaluation of Public Expenditures: The PPB System* vol. 1, 91st Cong., 1st sess. (Washington, D.C., Government Printing Office).

————, and Otto Eckstein. 1958. *Multiple Purpose River Development: Studies in Applied Economic Analysis* (Baltimore, Md., Johns Hopkins Press for Resources for the Future).

McKean, Roland. 1958. *Efficiency in Government Through Systems Analysis* (New York, Wiley).

U.S. Army Corps of Engineers. 1968. *Report on Replacement Locks and Dam No. 26, Mississippi River, Alton, Illinois, New 1200-Front Lock* (St. Louis, Mo.).

————. 1972. *Locks and Dam No. 26 (Replacement), General Design Memorandum* (St. Louis, Mo.).

————. 1975. *Locks and Dam No. 26 (Replacement), Formulation Evaluation Report* (St. Louis, Mo.).

————. 1976a. *Locks and Dam No. 26 (Replacement), Mississippi River, Alton, Illinois, Report of the Board of Engineers for Rivers and Harbors* (Fort Belvoir, Va.).

————. 1976b. *Locks and Dam No. 26 (Replacement), Revised Draft Supplement Environmental Statement* (St. Louis, Mo., January).

U.S. Congress, Joint Economic Committee. 1969. *The Analysis and Evaluation of Public Expenditures: The PPB System*, 91st Cong., 1st sess.

APPENDIX

"CONSERVATION
RECONSIDERED"
by
JOHN V. KRUTILLA

◆

INTRODUCTORY NOTE

Written as his plan of research and justification for establishing the Natural Environments Program at Resources for the Future in the mid-sixties, John V. Krutilla's "Conservation Reconsidered," reprinted in this appendix, is one of the most widely cited papers in modern resource and environmental economics. With more than two hundred citations since its publication in 1967, by all standards it is a seminal contribution to the field. Moreover, this total does not adequately reflect the paper's durability. Nearly 20 percent of these citations were made between 1983 and 1987. To appreciate the significance of this record, one must consider the average performance of journal articles in economics, as described by Michael C. Lovell in 1973.* He notes that of the articles cited in 1965 in four of the major journals, less than 8 percent were twenty years old or older.** It is also interesting that a large number of the paper's citations come from science-oriented journals. Thus, its influence has extended beyond economics.

Without mathematics or complex statistical analysis, "Conservation Reconsidered" not only outlined a decade's worth of systematic research but anticipated a wide range of conceptual problems and their solutions. These issues are increasingly recognized today as national and global concerns.

Most of the papers in this volume find their intellectual roots with this paper or the research it has stimulated. Therefore, it is particularly appropriate to have the paper reprinted here for new generations of resource economists to appreciate.

Thanks are due to the American Economic Association for allowing it to be reprinted in this volume.

V. Kerry Smith

---

*Michael C. Lovell, "The Production of Economic Literature: An Interpretation," *Journal of Economic Literature* vol. II (March 1973) pp. 27–55.

**The four journals considered in his analysis were the *American Economic Review*, *Econometrica*, the *Journal of Political Economy*, and the *Quarterly Journal of Economics*. Only *Econometrica* had a higher citation rate to older papers, with 12.2 percent to articles twenty years old or older.

◆

CONSERVATION RECONSIDERED

by John V. Krutilla*

Reprinted, with permission, from *American Economic Review* vol. 57 (September 1967) pp. 777–786.

"It is the clear duty of Government, which is the trustee for unborn generations as well as for its present citizens, to watch over, and if need be, by legislative enactment, to defend, the exhaustible natural resources of the country from rash and reckless spoliation. How far it should itself, either out of taxes, or out of State loans, or by the device of guaranteed interest, press resources into undertakings from which the business community, if left to itself, would hold aloof, is a more difficult problem. Plainly, if we assume adequate competence on the part of governments, there is a valid case for *some* artificial encouragement to investment, particularly to investments the return from which will only begin to appear after the lapse of many years."

A. C. Pigou

Conservation of natural resources has meant different things to different people. But to the economist from the time of Pigou, who first took notice of the economics of conservation [10, p. 27ff], until quite recently, the central concerns have been associated with the question of the optimal intertemporal utilization of the fixed natural resource stocks. The gnawing anxiety provoked by the Malthusian thesis of natural resource scarcity was in no way allayed by the rates of consumption of natural resource stocks during two world wars occurring between the first and fourth editions of Pigou's famous work. In the United States, a presidential commission, reviewing the materials situation following World War II, concluded that an end had come to the historic decline in the cost of natural resource commodities [12, pp. 13–14]. This conclusion reinforced the concern of many that the resource base ultimately would be depleted.

*The author is indebted to all of his colleagues at Resources for the Future and to Harold Barnett, Paul Davidson, Otto Davis, Chandler Morse, Peter Pearse, and Ralph Turvey for many helpful suggestions on an earlier draft of this paper.

263

More recently, on the other hand, a systematic analysis of the trends in prices of natural resource commodities did not reveal any permanent interruption in the decline relative to commodities and services in general [11]. Moreover, a rather ambitious attempt to test rigorously the thesis of natural resource scarcity suggested instead that technological progress had compensated quite adequately for the depletion of the higher quality natural resource stocks [1]. Further, given the present state of the arts, future advances need not be fortuitous occurrences; rather the rate of advance can be influenced by investment in research and development. Indeed, those who take an optimistic view would hold that the modern industrial economy is winning its independence from the traditional natural resources sector to a remarkable degree. Ultimately, the raw material inputs to industrial production may be only mass and energy [1, p. 238].[1]

While such optimistic conclusions were being reached, they were nevertheless accompanied by a caveat that, while we may expect production of goods and services to increase without interruption, the level of living may not necessarily be improved. More specifically, Barnett and Morse concluded that the quality of the physical environment—the landscape, water, and atmospheric quality—was deteriorating.

These conclusions suggest that on the one hand the traditional concerns of conservation economics—the husbanding of natural resource stocks for the use of future generations—may now be outmoded by advances in technology. On the other hand, the central issue seems to be the problem of providing for the present and future the amenities associated with unspoiled natural environments, for which the market fails to make adequate provision. While this appears to be the implication of recent research,[2] and is certainly consistent with recent public policy in regard to preserving natural environments, the traditional economic rationale for conservation does not address itself to this issue directly.[3] The use of Pigou's social time preference may serve only to hasten the conversion of natural environments into low-yield capital investments.[4] On what basis, then, can we make decisions when we confront a

---

[1]The conclusions were based on data relevant to the U.S. economy. While they may be pertinent to Western Europe also, all of my subsequent observations are restricted to the United States.

[2]For example, see [7].

[3]It must be acknowledged that with sufficient patience and perception nearly all of the argument for preserving unique phenomena of nature can be found in the classic on conservation economics by Ciriacy-Wantrup [3].

[4]An example of this was the recent threat to the Grand Canyon by the proposed Bridge and Marble Canyon dams. Scott makes a similar point with reference to natural resource commodities [13].

choice entailing action which will have an irreversible adverse consequence for rare phenomena of nature? I investigate this question below.

Let us consider an area with some unique attribute of nature—a geomorphologic feature such as the Grand Canyon, a threatened species, or an entire ecosystem or biotic community essential to the survival of the threatened species.[5] Let us assume further that the area can be used for certain recreation and/or scientific research activities which would be compatible with the preservation of the natural environment, or for extractive activities such as logging or hydraulic mining, which would have adverse consequences for scenic landscapes and wildlife habitat.

A private resource owner would consider the discounted net income stream from the alternative uses and select the use which would hold prospects for the highest present net value. If the use which promises the highest present net value is incompatible with preserving the environment in its natural state, does it necessarily follow that the market will allocate the resources efficiently? There are several reasons why private and social returns in this case are likely to diverge significantly.

Consider the problem first in its static aspects. By assumption, the resources used in a manner compatible with preserving the natural environment have no close substitutes; on the other hand, alternative sources of supply of natural resource commodities are available.[6] Under the circumstances and given the practical obstacles to perfectly discriminating pricing, the private resource owner would not be able to appropriate in gate receipts the entire social value of the resources when used in a manner compatible with preserving the natural state. Thus the present values of his expected net revenues are not comparable as between the competing uses in evaluating the efficiency of the resource allocation.

Aside from the practical problem of implementing a perfectly discriminating pricing policy, it is not clear even on theoretic grounds that a comparison of the total area under the demand curve on the one hand and market receipts on the other will yield an unambiguous answer to the allocative question.

---

[5]Uniqueness need not be absolute for the following arguments to hold. It may be, like Dupuit's bridge, a good with no adequate substitutes in the "natural" market area of its principal clientele, while possibly being replicated in other market areas to which the clientele in question has no access for all practical purposes.

[6]The asymmetry in the relation posited is realistic. The historic decline in cost of natural resource commodities relative to commodities in general suggests that the production and exchange of the former occur under fairly competitive conditions. On the other hand, increasing congestion at parks, such as Yellowstone, Yosemite, and Grand Canyon, suggests there are no adequate substitutes for these rare natural environments.

When the existence of a grand scenic wonder or a unique and fragile ecosystem is involved, its preservation and continued availability are a significant part of the real income of many individuals.[7] Under the conditions postulated, the area under the demand curve, which represents a maximum willingness to pay, may be significantly less than the minimum which would be required to compensate such individuals were they to be deprived in perpetuity of the opportunity to continue enjoying the natural phenomenon in question. Accordingly, it is conceivable that the potential losers cannot influence the decision in their favor by their aggregate willingness to pay, yet the resource owner may not be able to compensate the losers out of the receipts from the alternative use of the resource. In such cases—and they are more likely encountered in this area—it is impossible to determine whether the market allocation is efficient or inefficient.

Another reason for questioning the allocative efficiency of the market for the case in hand has been recognized only more recently. This involves the notion of *option demand* [14]. This demand is characterized as a willingness to pay for retaining an option to use an area or facility that would be difficult or impossible to replace and for which no close substitute is available. Moreover, such a demand may exist even though there is no current intention to use the area or facility in question and the option may never be exercised. If an option value exists for rare or unique occurrences of nature, but there is no means by which a private resource owner can appropriate this value, the resulting resource allocation may be questioned.

Because options are traded on the market in connection with other economic values, one may ask why no market has developed where option value exists for the preservation of natural environments.[8] We need to consider briefly the nature of the value in question and the marketability of the option.

From a purely scientific viewpoint, much is yet to be learned in the earth and life sciences; preservation of the objects of study may be defended on these grounds, given the serendipity value of basic research. We know also that the natural biota represents our reservoir of germ plasm, which has economic value. For example, modern agriculture in advanced countries represents cultivation figuratively in a hot-house environment in which crops are protected against disease, pests, and drought by a variety of agricultural practices. The energy released from some of the genetic characteristics no longer required for survival under cultivated conditions is redirected toward greater productivity. Yet because of the instability introduced with progressive re-

---

[7]These would be the spiritual descendants of John Muir, the present members of the Sierra Club, the Wilderness Society, National Wildlife Federation, Audubon Society and others to whom the loss of a species or the disfigurement of a scenic area causes acute distress and a sense of genuine relative impoverishment.

[8]For a somewhat differently developed argument, see [6].

duction of biological diversity, a need occasionally arises for the reintroduction of some genetic characteristics lost in the past from domestic strains. It is from the natural biota that these can be obtained.

The value of botanical specimens for medicinal purposes also has been long, if not widely, recognized. Approximately half of the new drugs currently being developed are obtained from botanical specimens.[9] There is a traffic in medicinal plants which approximates a third of a billion dollars annually. Cortisone, digitalis, and heparin are among the better known of the myriad drugs which are derived from natural vegetation or zoological sources. Since only a relatively small part of the potential medicinal value of biological specimens has yet been realized, preserving the opportunity to examine all species among the natural biota for this purpose is a matter of considerable importance.

The option value may have only a sentimental basis in some instances. Consider the rallying to preserve the historical relic, "Old Ironsides."[10] There are many persons who obtain satisfaction from mere knowledge that part of wilderness North America remains even though they would be appalled by the prospect of being exposed to it. Subscriptions to World Wildlife Fund are of the same character. The funds are employed predominantly in an effort to save exotic species in remote areas of the world which few subscribers to the Fund ever hope to see. An option demand may exist therefore not only among persons currently and prospectively active in the market for the object of the demand, but among others who place a value on the mere existence of biological and/or geomorphological variety and its widespread distribution.[11]

If a genuine value for retaining an option in these respects exists, why has not a market developed? To some extent, and for certain purposes, it has. Where a small natural area in some locality in the United States is threatened, the property is often purchased by Nature Conservancy,[12] a private organization which raises funds through voluntary subscriptions.[13] But this market is grossly imperfect. First, the risk for private investors associated with absence of knowledge as to whether a particular ecosystem has special characteristics not widely

---

[9]For an interesting account of the use of plants for medicinal purposes, see [8].

[10]The presumption in favor of option value is applicable also to historic and cultural features; rare works of art, perhaps, being the most prominent of this class.

[11]The phenomenon discussed may have an exclusive sentimental basis, but if we consider the "bequest motivation" in economic behavior, discussed below, it may be explained by an interest in preserving an option for one's heirs to view or use the object in question.

[12]Not to be confused with a public agency of the same name in the United Kingdom.

[13]Subscriptions to World Wildlife Fund, the Wilderness Society, National Parks Association, etc. may be similar, but, of course, much of the effect these organizations have on the preservation of natural areas stems not from purchasing options, but from influencing public programs.

shared by others is enormous.[14] Moreover, to the extent that the natural environment will support basic scientific research which often has unanticipated practical results, the serendipity value may not be appropriable by those paying to preserve the options. But perhaps of greatest significance is that the preservation of the grand scenic wonders, threatened species, and the like involves comparatively large land tracts which are not of merely local interest. Thus, all of the problems of organizing a market for public goods arise. Potential purchasers of options may be expected to bide time in the expectation that others will meet the necessary cost, thus eliminating cost to themselves. Since the mere existence or preservation of the natural environment in question satisfies the demand, those who do not subscribe cannot be excluded except by the failure to enroll sufficient subscribers for its preservation.

Perhaps of equal significance to the presumption of market failure are some dynamic characteristics of the problem suggested by recent research. First, consider the consumption aspects of the problem. Davidson, Adams, and Seneca have recently advanced some interesting notions regarding the formation of demand that may be particularly relevant to our problem [5, p. 186].

When facilities are not readily available, skills will not be developed and, consequently, there may be little desire to participate in these activities. If facilities are made available, opportunities to acquire skill increase, and user demand tends to rise rapidly over time as individuals learn to enjoy these activities. Thus, participation in and enjoyment of water recreational activities by the present generation will stimulate future demand without diminishing the supply presently available. Learning-by-doing, to the extent it increases future demand, suggests an interaction between present and future demand functions, which will result in a public good externality, as present demand enters into the utility function of future users.

While this quotation refers to water-based recreation, it is likely to be more persuasive in connection with some other resource-based recreation activity. Its relevance for wilderness preservation is obvious. When we consider the remote backcountry landscape, or the wilderness scene as the object of experience and enjoyment, we recognize that utility from the experience depends predominantly upon the prior acquisition of technical skill and specialized knowledge. This, of course, must come from experience initially with less arduous or demanding activities. The more the present population is initiated into activities requiring similar but less advanced skills (e.g., car camping), the better prepared the future population will be to participate in the more exacting activities. Given the phenomenal rise of car camping, if this activity will spawn a disproportionate number of future back-packers, canoe cruisers, cross-country skiers, etc., the greater will be the induced demand for wild, primitive,

---

[14]The problem here is in part like a national lottery in which there exists a very small chance for a very large gain. Unlike a lottery, rather large sums at very large risk typically would be required.

and wilderness-related opportunities for indulging such interest. Admittedly, we know little about the demand for outdoor experiences which depend on unique phenomena of nature—its formation, stability, and probable course of development. These are important questions for research, results of which will have significant policy implications.

In regard to the production aspects of the "new conservation," we need to examine the implications of technological progress a little further. Earlier I suggested that the advances of technology have compensated for the depletion of the richer mineral deposits and, in a sense, for the superior stands of timber and tracts of arable land. On the other hand, there is likely to be an asymmetry in the implications of technological progress for the production of goods and services from the natural resource base, and the production of natural phenomena which give rise to utility without undergoing fabrication or other processing.[15] In fact, it is improbable that technology will advance to the point at which the grand geomorphologic wonders could be replicated, or extinct species resurrected. Nor is it obvious that fabricated replicas, were they even possible, would have a value equivalent to that of the originals. To a lesser extent, the landscape can be manufactured in a pleasing way with artistry and the larger earth-moving equipment of today's construction technology. Open pit mines may be refilled and the surroundings rehabilitated in a way to approximate the original conditions. But even here the undertaking cannot be accomplished without the cooperation of nature over a substantial period of time depending on the growth rate of the vegetal cover and the requirements of the native habitat.[16] Accordingly, while the supply of fabricated goods and commercial services may be capable of continuous expansion from a given resource base by reason of scientific discovery and mastery of technique, the supply of natural phenomena is virtually inelastic. That is, we may preserve the natural environment which remains to provide amenities of this sort for the future, but there are significant limitations on reproducing it in the future should we fail to preserve it.

If we consider the asymmetric implications of technology, we can conceive of a transformation function having along its vertical axis amenities derived directly from association with the natural environment and fabricated goods along the horizontal axis. Advances in technology would stretch the transformation function's terminus along the horizontal axis but not appreciably along the vertical. Accordingly, if we simply take the effect of technological progress over time, considering tastes as constant, the marginal trade-off between manufactured and natural amenities will progressively favor the latter. Natural

---

[15]I owe this point to a related observation, to my knowledge first made by Ciriacy-Wantrup [3, p. 47].

[16]That is, giving rise to option value for members of the present population.

environments will represent irreplaceable assets of appreciating value with the passage of time.

If we consider technology as constant, but consider a change in tastes progressively favoring amenities of the natural environment due to the learn-by-doing phenomenon, natural environments will similarly for this reason represent assets of appreciating value. If both influences are operative (changes in technology with asymmetric implications, and tastes), the appreciating value of natural environments will be compounded.

This leads to a final point which, while a static consideration, tends to have its real significance in conjunction with the effects of parametric shifts in tastes and technology. We are coming to realize that consumption-saving behavior is motivated by a desire to leave one's heirs an estate as well as by the utility to be obtained from consumption.[17] A bequest of maximum value would require an appropriate mix of public and private assets, and, equally, the appropriate mix of opportunities to enjoy amenities experienced directly from association with the natural environment along with readily producible goods. But the option to enjoy the grand scenic wonders for the bulk of the population depends upon their provision as public goods.

Several observations have been made which may now be summarized. The first is that, unlike resource allocation questions dealt with in conventional economic problems, there is a family of problems associated with the natural environment which involves the irreproducibility of unique phenomena of nature—or the irreversibility of some consequence inimical to human welfare. Second, it appears that the utility to individuals of direct association with natural environments may be increasing while the supply is not readily subject to enlargement by man. Third, the real cost of refraining from converting our remaining rare natural environments may not be very great. Moreover, with the continued advance in technology, more substitutes for conventional natural resources will be found for the industrial and agricultural sectors, liberating production from dependence on conventional sources of raw materials. Finally, if consumption-saving behavior is motivated also by the desire to leave an estate, some portion of the estate would need to be in assets which yield collective consumption goods of appreciating future value. For all of these reasons we are confronted with a problem not conventionally met in resource economics. The problem is of the following nature.

At any point in time characterized by a level of technology which is less advanced than at some future date, the conversion of the natural environment into industrially produced private goods has proceeded further than it would have with the more advanced future technology. Moreover, with the apparent increasing appreciation of direct contact with natural environments, the conversion will have proceeded further, for this reason as well, than it would have were the future

---

[17]See [2]; also [9].

composition of tastes to have prevailed. Given the irreversibility of converted natural environments, however, it will not be possible to achieve a level of well-being in the future that would have been possible had the conversion of natural environments been retarded. That this should be of concern to members of the present generation may be attributable to the bequest motivation in private economic behavior as much as to a sense of public responsibility.[18]

Accordingly, our problem is akin to the dynamic programming problem which requires a present action (which may violate conventional benefit-cost criteria) to be compatible with the attainment of future states of affairs. But we know little about the value that the instrumental variables may take. We have virtually no knowledge about the possible magnitude of the option demand. And we still have much to learn about the determinants of the growth in demand for outdoor recreation and about the quantitative significance of the asymmetry in the implications of technological advances for producing industrial goods on the one hand and natural environments on the other. Obviously, a great deal of research in these areas is necessary before we can hope to apply formal decision criteria comparable to current benefit-cost criteria. Fully useful results may be very long in coming; what then is a sensible way to proceed in the interim?

First, we need to consider what we need as a minimum reserve to avoid potentially grossly adverse consequences for human welfare. We may regard this as our scientific preserve of research materials required for advances in the life and earth sciences. While no careful evaluation of the size of this reserve has been undertaken by scientists, an educated guess has put the need in connection with terrestrial communities at about ten million acres for North America [4, p. 128]. Reservation of this amount of land—but a small fraction of one per cent of the total relevant area—is not likely to affect appreciably the supply or costs of material inputs to the manufacturing or agricultural sectors.

The size of the scientific preserve required for aquatic environments is still unknown. Only after there is developed an adequate system of classification of aquatic communities will it be possible to identify distinct environments, recognize the needed reservations, and, then, estimate the opportunity costs. Classification and identification of aquatic environments demand early research attention by natural scientists.

Finally, one might hope that the reservations for scientific purposes would also support the bulk of the outdoor recreation demands, or that substantial

---

[18]The rationale above differs from that of Stephen Marglin which is perhaps the most rigorous one relying on a sense of public responsibility and externalities to justify explicit provision for future generations. In this case also, my concern is with providing *collective consumption goods for the present and future*, whereas the traditional concern in conservation economics has been with provision of *private intermediate goods for the future*.

additional reservations for recreational purposes could be justified by the demand and implicit opportunity costs. Reservations for recreation, as well as for biotic communities, should include special or rare environments which can support esoteric tastes as well as the more common ones. This is a matter of some importance because outdoor recreation opportunities will be provided in large part by public bodies, and within the public sector there is a tendency to provide a homogenized recreation commodity oriented toward a common denominator. There is need to recognize, and make provision for, the widest range of outdoor recreation tastes, just as a well-functioning market would do. We need a policy and a mechanism to ensure that all natural areas peculiarly suited for specialized recreation uses receive consideration for such uses. A policy of this kind would be consistent both with maintaining the greatest biological diversity for scientific research and educational purposes and with providing the widest choice for consumers of outdoor recreation.

REFERENCES

1. H. J. BARNETT AND C. MORSE, *Scarcity and Growth: The Economics of Natural Resource Availability*. Baltimore 1963.

2. S. B. CHASE, JR., *Asset Prices in Economic Analysis*. Berkeley 1963.

3. S. V. CIRIACY-WANTRUP, *Resources Conservation*. Berkeley 1952.

4. F. DARLING AND J. P. MILTON, ed., *Future Environments of North America, Transformation of a Continent*. Garden City, N.Y. 1966.

5. P. DAVIDSON, F. G. ADAMS, AND J. SENECA, "The Social Value of Water Recreation Facilities Resulting from an Improvement in Water Quality: The Delaware Estuary," in A. V. Kneese and S. C. Smith, ed., *Water Research*, Baltimore 1966.

6. A. E. KAHN, "The Tyranny of Small Decisions: Market Failures, Imperfections, and the Limits of Economics," *Kyklos*, 1966, 19 (1), 23–47.

7. A. V. KNEESE, *The Economics of Regional Water Quality Management*. Baltimore 1964.

8. M. B. KREIG, *Green Medicine: The Search for Plants that Heal*. New York 1964.

9. F. MODIGLIANI AND R. BRUMBERG, "Utility Analysis and the Consumption Function: An Interpretation of Cross-Section Data," in K. K. Kurihara, ed., *Post-Keynesian Economics*, New Brunswick 1954.

10. A. C. PIGOU, *The Economics of Welfare*, 4th ed., London 1952.

11. N. POTTER AND F. T. CHRISTY, JR., *Trends in Natural Resources Commodities: Statistics of Prices, Output, Consumption, Foreign Trade, and Employment in the United States, 1870–1957*, Baltimore 1962.

12. The President's Materials Policy Commission, *Resources for Freedom, Foundation for Growth and Security*, Vol. I. Washington 1952.

13. A. D. SCOTT, *Natural Resources: The Economics of Conservation*. Toronto 1955.

14. B. A. WEISBROD, "Collective Consumption Services of Individual Consumption Goods," *Quart. Jour. Econ.*, Aug. 1964, 77, 71–77.

THE WORKS OF
JOHN V. KRUTILLA:
A BIBLIOGRAPHIC PROFILE

Rather than reporting a complete list of John V. Krutilla's research to date, this appendix adopts a strategy that he taught those of us who worked with him. Organize your research to meet long-term goals, stake out a big problem, and try to address it with an integrated sequence of discrete research projects. John's research career provides a model of this strategy. It seems appropriate, therefore, to try to capture the results of that focusing in research objectives with a profile of his work in five major areas. What follows is a selected (and therefore partial) listing of his research contributions. As of this writing John has written, edited, or contributed chapters to forty books and six monographs. He has also written or been coauthor of more than forty journal articles and numerous reports.—*V.K.S.*

I. Principles and Practices for Evaluating Public Investments in Water Resources

1958   With Otto Eckstein. *Multiple Purpose River Development: Studies in Applied Economic Analysis.* Baltimore, Md., Johns Hopkins Press for Resources for the Future.

         With Otto Eckstein. "The Cost of Federal Money, Hells Canyon and Economic Efficiency." Parts 1, 2. *National Tax Journal* vol. 11 (March, June) pp. 1–20, 114–128.

1961   "Welfare Aspects of Benefit–Cost Analysis." *Journal of Political Economy* vol. 69 (June) pp. 226–235.

         With Maynard Hufschmidt and Julius Margolis, and with the assistance of Stephen Marglin. *Standards and Criteria for Formulating and Evaluating Federal Water Resource Development.* Report of Panel of Consultants to the Bureau of the Budget. Washington, D.C.

I. *Continued*

1967   With Robert Haveman. "Unemployment, Excess Capacity, and Benefit–Cost Investment Criteria." *Review of Economics and Statistics* vol. 49 (August) pp. 382–392.

1968   With Robert H. Haveman, and with the assistance of Robert M. Steinberg. *Unemployment, Idle Capacity, and the Evaluation of Public Expenditures: National and Regional Analyses.* Baltimore, Md., Johns Hopkins Press for Resources for the Future.

II. The Role of Institutions in the Evaluation of Public Investments and the Management of Public Resources

1966   "Is Public Intervention in Water Resources Development Conducive to Economic Efficiency?" *Natural Resources Journal* vol. 6 (January) pp. 60–75.

1967   *The Columbia River Treaty: The Economics of an International River Basin Development.* Baltimore, Md., Johns Hopkins Press for Resources for the Future.

1969   "Efficiency Goals, Market Failure, and the Substitution of Public for Private Action," in *The Analysis and Evaluation of Public Expenditures: The PPB System.* U.S. Congress, Joint Economic Committee, 91st Cong., 1st Sess. Washington, D.C., Government Printing Office.

1975   "The Use of Economics in Project Evaluations." Pp. 374–381 in *Transactions of the 40th North American Wildlife and Natural Resource Conference.* Washington, D.C., Wildlife Management Institute.

III. The Economic Analysis of Investment and Management Decisions for Unique Natural Environments and Wildlife Resources

1967   "Conservation Reconsidered." *American Economic Review* vol. 57 (September) pp. 777–786.

1972   . . . , editor and contributing author. *Natural Environments: Studies in Theoretical and Applied Analysis.* Baltimore, Md., Johns Hopkins University Press for Resources for the Future.

1972 With Charles J. Cicchetti and Anthony C. Fisher. "The Economics
(cont.) of Environmental Preservation: A Theoretical and Empirical Analysis." *American Economic Review* vol. 62, pp. 605–619.

With Charles J. Cicchetti, A. Myrick Freeman III, and Clifford
S. Russell. "Observations on the Economics of Irreplaceable As-
sets," in A. V. Kneese and B. T. Bower, eds., *Environmental Quality
Analysis, Theory and Method in the Social Sciences.* Baltimore, Md.,
Johns Hopkins Press for Resources for the Future.

With Anthony C. Fisher. "Determination of Optimal Capacity
of Resource-Based Recreational Facilities," *Natural Resources Journal* vol. 12 (July) pp. 417–444.

1974 With Anthony C. Fisher. "Valuing Long Run Ecological Consequences and Irreversibilities." *Journal of Environmental Economics and Management* vol. 1 (August) pp. 96–108.

With Anthony C. Fisher and Charles J. Cicchetti. "Reflections
on Irreversibilities in Economic Process." *American Economic Review* vol. 64 (December) pp. 1030–1039.

1975 Anthony C. Fisher. *The Economics of Natural Environments: Studies
in the Valuation of Commodity and Amenity Resources.* Baltimore,
Md., Johns Hopkins University Press for Resources for the Future.

With Anthony C. Fisher. "Resource Conservation, Environmental
Preservation and the Rate of Discount." *Quarterly Journal of
Economics* vol. 89 (August) pp. 358–370.

IV. Resource and Environmental Constraints to Long-Term Economic Well-Being and Growth

1977 "Resource Availability, Environmental Constraints and the Ed-
ucation of a Forester." The Maughan Memorial Lecture, Duke
University. In F. J. Convery and J. E. Davis, eds., *Centers of
Influence and U.S. Forest Policy.* Durham, N.C., Duke University,
School of Forestry and Environmental Studies.

1979 With V. Kerry Smith. "Resource and Environmental Constraints
to Growth." *American Journal of Agricultural Economics* vol. 61
(August) pp. 395–408.

1982 . . . , and V. Kerry Smith, editors. *Explorations in Natural Resource
Economics.* Baltimore, Md., Johns Hopkins University Press for
Resources for the Future.

IV. *Continued*

1984 With V. Kerry Smith. "Economic Growth, Resource Availability, and Environmental Quality." *American Economic Review* vol. 74 (May) pp. 226–230.

V. The Economic Principles of Multiple Use Public Land Management

1978 With John A. Haigh. "An Integrated Approach to National Forest Management." *Environmental Law* vol. 8 (Spring) pp. 373–415.

1980 With John A. Haigh. "Clarifying Policy Directives: The Case of National Forest Management." *Policy Analysis* (Fall) pp. 375–384.

1983 With Anthony C. Fisher, William F. Hyde, and V. Kerry Smith. "Public Versus Private Ownership: The Federal Lands Case." *Journal of Policy Analysis and Management* vol. 2 (Summer) pp. 548–558.

1987 "Below Cost Sales: Tying up the Loose Ends of the National Forest Management Act." *Journal of Forestry* vol. 85 (August) pp. 27–29.

In press With Michael D. Bowes. *Multiple-Use Management: The Economics of Public Forestlands*. Washington, D.C., Resources for the Future.

INDEX

INDEX

Adams, D. M., 138–142, 145, 156,
  159 nn. 2, 4
Adams, F. Gerard, 50
Administrative Procedures Act of
  1946, 241
Agricultural development, habitat loss
  from, 60
Amemiya, Takeshi, 175
American Rivers Conservation
  Council, 192
Anderson, Terry L., 192
Andrus, Cecil, 223, 226, 229
Army Corps of Engineers, 192
  Locks and Dam No. 26 case and,
    247–251
  Reimbursement policy, 202
  Responsibility for inland navigation
    system, 245
Arrow, Kenneth J., 20, 21, 28, 62,
  68 n. 2, 72, 78, 81
Association of American Railroads,
  247
Association of Environmental and
  Resource Economists, 5
Atkinson, Anthony B., 131 n. 4
Auerbach, Alan J., 131 n. 4
Aumann, R. J., 36 n. 23

Baker, Howard H., Jr., and
  Endangered Species Act
  exemption, 217–218, 226–227,
  229
Ballard, Charles L., 127

Bartik, Timothy J., 170
Barzel, Yoram, 92
Baumol, William J., 5–6; 36 nn. 21,
  25, 27; 121; 131–132 n. 6
Bell, Clive, 32
Bell, Griffin, 215
Benefit–cost analysis of public
  policy, 5
  Applied to developing countries,
    33 n. 3
  Applied to natural environment
    development decisions, 11, 19,
    21–22, 60–62, 239–240
  Applied to water resources projects,
    46, 193
  Applied welfare economics and, 7,
    8, 28, 71
  Conceptual rationale for, 34 n. 7
  For intergenerational equity
    determination, 28, 79–82
  Techniques for, 193–194
Berck, P., 64
Berkman, Richard L., 192
Bevill, Tom, 229
Bishop, R. C., 15, 67 n. 1, 166
Blackwelder, Brent, 230
Bockstael, Nancy E., 176, 178
Bohm, Peter, 10
Boiteux, M., 120
Bös, Dieter, 36 n. 26, 131 nn. 1, 3
Bower, Blair T., 48
Bowes, Michael D., 9, 18, 27, 65,
  131 n. 6, 160 n. 6, 165, 185
Boyle, K. J., 15

Bradford, David F., 8, 121
Brewer, D., 171, 172, 176
Bromley, Daniel W., 24
Brookshire, David S., 19, 166, 193
Brown, F. Lee, 49
Brown, Gardner M., Jr., 11, 13, 91, 92, 166
Brown, James N., 170
Browning, Edgar K., 127
Budget Control Act of 1974, 240
Bureau of Reclamation, 52, 192
   Cost-sharing proposals, 203–204
   Reimbursement policy, 195, 196, 197, 201
Bureau of the Budget, 192
Bureau of the Census, 160 n. 8
Burness, H. S., 195
Burt, O. R., 171, 172, 176
Burton, Ian, 30

Cornes, Richard, 18
Carroll, J. L., 248, 258
Carson, Richard T., 19, 166
Carter, Jimmy
   Cost-sharing recommendations for water projects, 201
   Dilemma over Tellico project, 229–230
   And Endangered Species Act Amendments, 218
   Energy independence promotion, 227
Central Utah Water Conservancy District, 204
CERCLA. See Comprehensive Environmental Response, Compensation, and Liability Act of 1980
Cesario, F. J., 171
Cicchetti, Charles J., 11, 18, 19, 22, 25, 34 n. 9, 35 n. 20, 166, 171, 234 n. 17, 237, 258

Ciriacy-Wantrup, S. V., 14, 29, 91, 116
Clark, C. W., 64
Clawson, Marion, 50, 171
Colorado River
   Big Thompson project, 198, 200
   Compact for upper basin development, 52
   Law of the River applied to, 45
Columbia River Basin development, 47–48
Comprehensive Environmental Response, Compensation, and Liability Act (CERCLA) of 1980, 31
Congressional Budget Office, 197, 201, 240
"Conservation Reconsidered," Krutilla, 4, 11, 14, 51, 59, 72, 91, 211, 263–273
Cook, Phillip J., 15, 30, 34, 34 n. 13
Council on Environmental Quality, 201
Court, Louis M., 167
Cropper, Maureen L., 11
Culver, John C., 217, 226
Cummings, Ronald G., 19, 193

Daniels, Steven E., 22, 24, 25, 35 n. 15, 135
d'Arge, Ralph C., 11
Dasgupta, Partha S., 82, 114
Davidson, Paul, 50
Davis, Robert K., 22, 23, 50, 166, 211
Deacon, Robert T., 192
DeHaven, James C., 120, 131 n. 5
Desvousges, William, 32, 186 n. 3
Developing countries, benefit–cost analysis applied to, 33 n. 3
Dickinson, Robert E., 131 n. 5
Discount rate, public investment, 6
   Concepts, 84

Intergenerational efficiency and, 73, 84

Intergenerationally fair, 20, 76; aggregation rule for, 77–79, 80, 82–83, 85, 86; asymmetric rate for, 80; stationarity axiom, 82, 85; transitivity rule, 80

Levels, 82–83, 88

Opportunity cost criteria for, 20, 83

Domenici, Pete V., 234 n. 19

Dorfman, Robert and Nancy S., 193

Drew, Elizabeth, 217, 234 n. 10

Dreze, Jean, 33 n. 3

Drucker, Peter, 206–207

Duncan, John T., 226, 230

Easter, K. William, 192

Eckstein, Otto, 4; 10; 14; 19; 20; 45; 46; 47; 119; 120–121; 131 nn. 4, 5; 135; 159; 192; 222; 258 n. 1

Edelson, N., 17

Efficiency, economic
Benefit–cost analysis and, 193
Integration of equity and, 71–72
Intergenerational, 72–73, 84, 86
Pareto, 84, 193
Resource allocation for, 4, 6; defined, 7; marginal conditions for, 8; optimal taxes for, 10
Resource investment and management, 22, 26, 35 n. 15; for recreation resources, 148, 156

Ehrlich, A. H. and P. R., 66

Eisenhower administration, private role in river basin development, 4

Electric utilities
Environmental problems related to, 238

Pricing policy: steps in case proceedings on, 252–253; state rules for, 251–252
See also Madison Gas & Electric Company case

Elsner, G., 147

Endangered species
Economic analysis of extinction, 11–12; from habitat loss, 59, 60–63; from overexploitation, 64–67
Noneconomic approach to, 67
Valuing of, 12, 91–92

Endangered Species Act of 1973, 23
Amendments, 1978, 216–218, 226–227, 230
Snail darter and implementation of, 23, 212
Tellico project and, 212, 214, 230; arguments against exemption, 227–228; court decision on, 214–215

Endangered Species Committee, 217, 223, 226, 231, 232

Energy and Water Development Appropriations Bill, 229–230

Environment
Decisions relating to: benefit–cost analysis for, 240; executive branch general rules for, 241–242
Ethical principles for, 29
Government role in control on, 237–238
Litigation on. See Electric utilities, pricing policy; Locks and Dam No. 26 case; Madison Gas & Electric case; Navigation projects
Measuring benefits of improvements in, 50
Rationality as basis for decisions on, 239, 240–242; court rules for, 242–243, 257, 258; government role in, 244–245, 257

Environment (*cont.*)

Regulation of: benefit–cost analysis
for, 8, 18, 30–31, 239;
effectiveness, 238, 239

See also Natural environment
resources

Environmental Defense Fund, 247,
253

Environmental Protection Agency
(EPA), 18, 50

Equality, equity versus, 74–75

Equity

Equality versus, 74–75

Integration of efficiency and, 71–72

Intergenerational: analogy with
intragenerational equity, 73–74,
80–81; asymmetry problem, 75–
76; capital productivity, 76, 83,
88; described, 73; discount rate
and, 20; distributional effects, 75,
76; externalities and, 74;
implementation problem in, 76;
individual preferences and, 75,
82–83; time preference and, 76,
83

Etnier, David, 214

Eubanks, L. S., 166

Externalities

Defined, 9–10

Equity and, 74

Intergenerational versus
intertemporal, 10–11

Means of dealing with, 6, 33 n. 1

Potential importance, 10

R&D activities and, 13

Federal Interagency River Basin
Committee, 46

Feldman, Allan M., 79, 80

Ferejohn, John, 20, 82, 85

Fischhoff, Baruch, 30

Fish and Wildlife Service, 214

Fisher, Anthony C., 5, 11, 13, 16, 19,
20–21, 29, 34 n. 9, 59, 68 n. 2,
158, 165, 166, 167, 211, 231,
232

Fisher, F. M., 167

Fishing (recreation), 167, 170

Fitzgerald, John, 131

Fitzgerald, Michael R., 216, 230,
233 n. 7

Fleming, J. M., 120, 121

Flood Control Act of *1936*, 46, 245

Flood control projects

Cost-sharing policy, 202

Valuing policies affecting, 29–30,
49

Forest lands

Harvesting practices: criteria for,
136–137; "even-flow"
requirement on, 24–25, 137;
market-sensitive criteria for, by
region, 142–147; production,
137, 156; public versus private,
145; welfare effects, 142, 145,
146, 147

Management, 137

Model, 138; public inventory, 141–
142; regional analysis, 142–147;
stumpage demand, 138–139;
143; stumpage supply, 139–140,
143

Multiple use; management cost
allocation problem, 27; value of
unharvested stock, 65–66

Forest Service, U.S., 141, 148

"Even flow" requirement on
harvesting practices, 24–25, 137

Forest inventory measures, 141–142

Harvesting criteria, 136–137

Program Accounting and
Management Attainment
Reporting System, 149

Timber management budget, *1977*,
138, 159 n. 1

Fox, Irving K., 48
Franklin, Douglas R., 196, 201
Freeman, A. Myrick, III, 15, 17, 18,
    22, 26, 119
Freeman, S. David, 215–216

Gallatin, Albert, 46
General Accounting Office (GAO),
    203, 204, 213, 240
    Review of six irrigation projects,
    197, 198
Genetic resources, valuation of stock
    of, 12, 13, 92–98, 114–115
Gillen, W., 258
Goldstein, Jon H., 92
Graff, Thomas J., 204
Graham, Daniel A., 16, 30, 34 n. 13
"Greenbook," economic analysis of
    water resources projects, 46
Greenwalt, Lynn A., 233 n. 5
Griliches, Zvi, 116 n. 4
Groundwater
    Contamination, 52–53
    Depletion effect on economic
    growth, 49

Hageman, Ronda K., 196, 201
Haigh, John M., 165
Hammack, Judd, 166
Hanemann, W. Michael, 15, 29, 63,
    68 n. 2, 175, 176, 177, 178
Hansson, Bengt, 85
Harrington, Winston, 24
Harsanyi, John, 76
Hartman, Richard C., 16, 34 n. 13, 65
Harvard Water Program, 47–48
Hashimoto, T., 131 n. 5
Hausman, J., 186 n. 2
Haveman, Robert H., 4; 10; 17; 23;
    25; 32; 35 nn. 19, 20; 128; 159;
    237; 246; 258
Haynes, R. W., 138–142, 159 n. 2

Heal, Geoffrey M., 82
Heaney, James P., 131 n. 5
Heberlein, T. A., 166
Hechman, J. J., 175
Hells Canyon case, 211, 234 n. 17
    Tellico case compared with, 212,
    231–232
Hells Canyon reach, 34 n. 9, 47
Henry, C., 68 n. 2
Herfindahl, Orris C., 30, 36 n. 28
Hicks, J. R. See Kaldor–Hicks
    concepts
High Plains Associates, 49
Hill, Hiram G., Jr., 214
Hirshleifer, Jack, 120, 131 n. 5
Hodge, Clarence Lewis, 230
Holdren, J. P., 66
Hoover Dam, 204
Hotelling, Harold, 171
Houthakker, H. S., 167
Howe, Charles W., 23, 26, 191, 192
Hueth, Darrell L., 33 n. 5
Hufschmidt, Maynard, 120, 192, 240
Hunting (recreation), 18–19, 167
    Hedonic model applied to, 170
    Simple repackaging model for, 178–
    183
Hyde, William F., 22, 24, 25, 35 n. 15,
    135, 145
Hydroelectric power, 10
    Cost sharing for, 203
    Pricing for Hoover Dam, 204
    Reimbursement of costs for, 195
    See also Hells Canyon case

Inland Waterways Trust Fund, 202
Intergenerational equity. See Equity,
    intergenerational
Interior, Department of the, 222, 230
    Joint study with TVA on Tellico
    project, 215–216
    Office of Policy Analysis, 218
Investment. See Public investment

Irrigation projects, reimbursement
    policy, 196, 197–198

Jackson, David H., 159 n. 3
James, L. Douglas, 120
Johnson, F. Reed, 30
Johnson, Lyndon B., 206
Johnson, M. Bruce, 17, 192
Johnston, J. Bennett, Jr., 227
Just, Richard E., 33 n. 5

Kaldor, Nicholas. See Kaldor–Hicks
    concepts
Kaldor–Hicks concepts
    Potential compensation test, 7,
        33 n. 5
    Potential Pareto improvement
        criteria, 79–80
Karaganis, J. V., 258
Karni, Edi, 16
Kates, Robert W., 30, 50
Kealy, M. J., 166
Kelly, Robert A., 49
Kelso, Maurice M., 192
Kincaid, Eugene, 233 n. 4
Kincaid, J., 159 n. 4
Kling, Catherine, L., 175, 178
Kneese, Allen V., 8, 12, 16, 18, 29,
    30, 45, 48, 49, 50, 73, 211, 239
Knetsch, Jack L., 50, 165, 219
Knutson–Vandenberg Act, 1930, 138
Kogiku, K. C., 11
Koopmans, Tjalling, 21, 72, 81–82
Kopp, Raymond J., 36 n. 29
Krupnick, Alan J., 24
Krutilla, John V., 3, 45, 65, 67 n. 1,
    114
    Analysis of environmental resources
        and policy measures, 22, 239,
        258 n. 1
    Analysis of water resources policies
        and projects, 9, 191–192, 193,
        211, 231, 232, 234 n. 17

On applied welfare economics, 5–6,
    10
Behavioral research on recreation
    resources use, 17
"Conservation Reconsidered," 4,
    11, 14, 51, 59, 72, 91, 211, 263–
    273
Economic analysis of Columbia
    Basin international treaty, 47
Harvard Water Program and, 47
On intergenerational conflicts of
    interest, 72–73
Pricing policies for natural resource
    outputs, 27, 120, 121, 131
On public land management
    practices, 18, 22, 24, 25–26,
    135, 158–159, 160 n. 6
On resource allocation problems,
    4–5, 91, 115, 165, 166, 185
On resource valuation, 14–16, 19–
    20
Kurz, Mordecai, 78

Lee, Richard R., 120
Legislation, environmental
    Interests influencing, 24
    For management responsibilities,
        21, 22
    Phased implementation of, 18
Leman, Christopher K., 233
Leopold, Aldo, 91
Lichtenstein, Sarah, 30
Lind, Robert, 20, 35 n. 17, 62
Locks and Dam No. 26 case
    Benefit–cost studies for, 248–250,
        259 n. 4
    Court decision for government,
        1979, 250
    Court decision for plaintiffs, 248
    Described, 247
    Legislation resulting from, 250–251
Loehman, Edna, 36

Loomis, J. B., 154
Loughlin, James C., 131 n. 5
Lucas, Robert C., 17, 18

McCloskey, Donald N., 233
McConnell, Kenneth E., 15
McGivney, M. P., 186 n. 3
McKean, Roland, 259 n. 1
Maass, Arthur, 7, 47
Mack, Lawrence E., 192
Madison Gas & Electric Company case
    Described, 253
    Environmentalists' arguments, 254,
        257–258
    Reforms resulting from, 256–257
    Role of economic analysis in,
        35 n. 19, 255–256
    Side effects, 256
    Utilities' arguments, 253
Magat, Wesley A., 24
Mäler, Karl Göran, 15, 171
Management. See Public management
    of environmental resources;
    Recreation resources,
    management
Manne, A. S., 120, 121
Mansfield, E., 111, 117 nn. 4, 5
Marglin, Stephen A., 92, 120
Martin, William E., 192
May, Carol, 258
Mendelsohn, Robert, 170–171
Milliman, Jerome W., 120, 131 n. 5
Mitchell, Robert Cameron, 19, 166
Models
    Forest land sector econometric,
        138–147
    Genetic resources R&D, 92–95,
        114–115
    Habitat loss, 60–63
    Recreation, 18–19, 166; hedonic
        site-choice, 168–171, 174, 176;
        multiple-site travel-cost, 171–
        174, 175, 186 n. 4; simple

repackaging, 178–184; 186 nn. 1,
    3; site-choice multinomial logit,
    176–178
Resources overexploitation, 64–67
Morey, E. R., 176
Muellbauer, J., 167, 186 n. 1
Municipal–industrial water supply,
    cost sharing for, 203
Muth, Richard F., 167
Myers, John T., 226
Myers, Norman, 91

National Association of Regulatory
    Utility Commissions, 256
National Environmental Policy Act
    (NEPA), 213, 230, 245, 246
National Park Service, 148
National Science Foundation, 255
National Water Commission, 192
National Wilderness Preservation
    System, 232
Natural environment resources
    Decisions involving development:
        benefit–cost analysis for, 11; bias
        in favor of, 63; preservation
        versus, 12, 60–61; risk and, 62–
        63; uncertainty in, 11, 12, 20,
        61, 63
    Krutilla's allocation decisions
        involving, 4–5
    Management, 6, 8–9, 10, 21–27
    Nonrenewable, 34 n. 10
    Valuation of nonmarketed, 13–14;
        existence, 15; models of, 18–19;
        quality, 17–18; survey techniques
        for, 19; under uncertainty, 15–
        16; uniqueness and, 16, 34 n. 13;
        use versus nonuse, 14–15; utility
        function and, 14, 34 n. 12,
        35 n. 16
    See also Water resources projects
Natural Environments Program, RFF,
    17, 19, 59, 72

Navigation projects, 25
  Army Corps of Engineers and, 245
  Challenges to, 247
  Procedures for evaluating
    environmental effects, 246
  Rates charged for, 246–247
  Reimbursement policy, 196
  See also Locks and Dam No. 26 case
Neely, Walter P., 196
Nelson, Gaylord A., 217, 234 n. 10
Nelson, Robert H., 233
NEPA. *See* National Environmental
    Policy Act
Norris, Jeffrey C., 22, 26, 119, 131
North, Ronald M., 96
Northern Colorado Water
    Conservancy District, costs, 198,
    200

O'Connor, Charles J., 30
Office of Technology Assessment, 52
Ogallala aquifer region, 49
Okada, N., 131 n. 5
Oldfield, Margery L., 91
Opportunity costs
  And discount rate choice, 19–20,
    83
  Dynamic, 4
  Endangered species management,
    158
  For resources preservation, 60
Option value, 15, 16, 34 n. 13, 60
Organisation for Economic
    Cooperation and Development,
    83

Page, Talbot, 10, 21, 28, 29, 71, 82,
    85
Pareto efficiency, 84, 193
Pareto improvement
  Actual, 74, 75
  Benefit–cost analysis for, 34 n. 7

Potential: decisions based on, 74,
    75, 88; dependence on
    redistribution, 79; Scitovsky
    amendment to, 80
Pareto optimality. *See* Efficiency
Parfit, Derek, 75
Pigou, A. C., 73, 76
Plater, Zygmunt J. B., 215, 235 n. 23
Plummer, Mark L., 16, 34 n. 13
Porter, Richard C., 11
Potential compensation test, 7, 33 n. 5
Powell, Lewis F., Jr., 215
Prewitt, R. E., 171
Prices
  Electric power: case proceedings
    on, 35 n. 19, 252–253; state
    rules for, 251–252
  Shadow, 4, 33 n. 3, 65
  Water resources projects: for
    efficiency versus cost recovery,
    119–121; Ramsey–Baumol–
    Bradford rules, 121–130
Public investment
  Benefit–cost evaluation of, 19, 21
  Efficiency criteria for, 8, 9, 35 n. 15
  Socially optimal, 11, 12
  Uncertainty in, 11–12, 20
  See also Discount rate, public
    investment
Public lands
  Management: effect on recreational
    lands' capacity and quality, 18;
    interaction between public
    investment decisions and, 22
  Multiple use: cost allocation
    problems, 27; criteria for, 9, 18;
    for maximization of social
    benefits, 165
  See also Forest lands; Public
    management of environmental
    resources
Public management of environmental
    resources, 6

Efficiency criteria for, 8–9, 35 n. 15
Externalities, 10
Implementation problems in, 25
Legislation for, 21–22
Multiple use criteria for, 9, 18
Political rules for, 23
Pricing rules for, 25–27
Public Utility Regulatory Policies Act
   of *1978*, 256

Quality of the Environment Program,
   RFF, 49

Ramsey, Frank P., 73, 76, 121
Ramsey–Baumol–Bradford (RBB)
   pricing rules, 25–26; 36 nn. 21,
   22; 130. *See also* Water resources
   projects, multiple purpose
Randall, Alan, 14, 34 n. 7, 166
Rao, S., 248
RBB pricing rules. *See* Ramsey–
   Baumol–Bradford pricing rules
Reagan, Ronald, on benefit–cost
   analysis for new regulations, 33–
   34 n. 6, 201
Rechichar, Steven J., 216, 230,
   233 n.7
Reclamation Act of *1902*, 46, 51
Recreation resources
   Consumer site choice, 166;
      multinomial logit model for,
      176–178; probability in, 176
   Demand–supply analysis, 157
   Hedonic site-choice model: applied
      to fishing and hunting, 170;
      evaluation, 171, 184;
      identification assumption, 169–
      170, 184; multiple-quantity unit
      assumption, 168–169; no-
      repackaging assumption, 167–
      168; stochastic elements, 174–
      175; travel-cost model versus,
      174, 176

Management: efficiency criteria for,
   136; resources allocation, 147–
   148; role of models in providing
   knowledge for, 185; welfare
   effects, 158
Methods of valuing, 18–19
Multiple-site travel-cost model,
   171–174, 186 n. 3; limited
   dependent variable problem, 175,
   186 n. 4; stochastic element in,
   174–175
Simple repackaging model: utility
   function and constraints, 178–
   180, 187 n. 6; visits as quantity
   units, 180–184
Welfare gains from improving
   quality of, 166, 171, 176
Wilderness, 9, 17–18
See also Seeley–Swan campground,
   Montana
Regulations
   Environmental: for damage
      assessment, 31; distributional
      effects, 8; implementation, 18,
      25; risk management in social,
      31
   Utilities, 238
Research and development (R&D), 12
   Appropriability of benefits from, 92
   Expenditures, 95–97; Nash duopoly
      in, 106–113
   Externalities from, 13
   Model, 92; demand
      interdependence in, 94;
      described, 94–95; with
      monopolies, 93–94, 95–96, 114–
      115, 116 n. 1; multimarket, 97–
      98
   Optimum subsidy level, 93, 96–97;
      investment tax credit for, 95,
      103; to maximize social benefits,
      94; private and social, 100–102;
      productivity, 106; *(cont.)*

Research and development (R&D)
    Optimum subsidy level (*cont.*)
        uniform versus, 111–113; welfare
        level, 101, 102
    Production function, 98–100
    Spillovers of, 92–93, 95, 100, 103–
        104, 113–114
    Subsidization for investment in:
        elasticity of demand and, 104,
        109–111; investment tax credit
        for, 95, 103
    Underinvestment in, 13
    Welfare levels, 100–102
Resources for the Future
    Natural Environments Program, 17,
        19, 59
    Quality of the Environment
        Program, 49
    Water Quality Program, 12, 48
Risk management in environmental
    development, 30–31, 62–63
River basin development, 4, 47–48
Rivers and Harbors and Flood Control
    Act of *1970*, 245–246
Rocky Mountain region
    Nonmarket forest value, 147
    Public versus private timber
        harvests, 142, 145
Roosevelt, Franklin D., 230
Rosen, Sherwin, 167, 168, 170,
    185 n. 1
Russell, Clifford S., 10, 32, 49,
    186 n. 3

Sandler, Todd, 18
SARA. *See* Superfund Amendments
    and Reauthorization Act of *1986*
Schmitz, Andrew, 33 n. 5
Schultze, Charles, position on Tellico
    project, 223, 226, 229, 231,
    234 n. 18

Schulze, William D., 19, 29, 193
Scitovsky, Tibor, 33 nn. 1, 5
Seckler, David, 206–207
Seeber, Lynn, 233 n. 5
Seeley–Swan campground, Montana,
    25, 136
    Costs: fixed versus variable, 149–
        151, 153; maintenance, 151–152;
        replacement, 152; vehicle use,
        153, 154
    Management: compliance checks,
        152–153; law enforcement, 153;
        resources allocation, 148; user
        fees, 153, 155
    Supply and demand analysis, 154–
        156, 157–158, 160 n. 8
Seneca, Joseph, 50
Shabman, Leonard, 205, 233,
    234 n. 15
Shadow prices, 33 n. 3
    Determination, 4
    Exploitation model, 65
Shapley, L. S., 36 n. 23
Shechter, Mordechai, 18
Shell, K., 167
Shoven, John B., 127
Slovic, Paul, 30
Smith, Adam, 71
Smith, V. Kerry, 3; 15; 16; 18; 19; 30;
    34 nn. 6, 13; 36 n. 29; 67 n. 1;
    131; 160 n. 9; 166; 170; 171;
    186 n. 3; 258
Smolensky, P., 258
Snail darter, 23
    Endangered Species Act and, 212
    Survival, 231, 235 n. 24
    Tellico Dam and Reservoir Project
        and, 213, 214, 232
Social choice, intergenerational
    Axioms, 78: Arrow, 81, 84–85, 87–
        88; Koopmans, 81–82, 86–87,
        88; Pareto, 78

Benefit–cost analysis for, 80
Correct discount rate for, 73, 76,
   83–84, 88
Decision rules for: aggregation, 78–
   79, 80, 82, 83, 85, 86;
   asymmetry, 80; constraints on,
   87–88; majority, 80, 85, 88;
   transitivity, 80, 84
And discounting levels, 82–83
Distributional effects, 76, 79
Language, 77, 81
Soil Conservation Service, 197
Solow, Robert, 76
Sorg, C. F., 154
Species
   Economic value, 12, 91–92
   Preservation, 91
   Private versus public demand for,
   92
Species extinction
   Economic versus noneconomic
   approach to, 67
   Habitat loss and: causes, 59–60;
   model, 60–63; prevention, 60
   Market structure flaws and, 113
   Overexploitation and: rate of
   harvest and, 64; model, 64–67
   Species variation to prevent, 66
Spence, A. Michael, 13, 92, 93, 94,
   114
Spofford, Walter O., Jr., 49
Squire, Lyn, 193
Stankey, George H., 17
Stennis, John C., 217
Stern, Nicholas, 33 n. 3
Stiglitz, Joseph E., 34 n. 10, 114,
   131 n. 4
Stoll, John R., 14
Superfund Amendments and
   Reauthorization Act (SARA) of
   1986, 31
Swierzbinski, Joseph, 11, 13, 91, 92

Tellico Dam and Reservoir case, 211
   And completion of dam, 226, 229,
   232
   Congressional vote on, 228, 229
   Described, 213
   Economic analysis for, 219–222
   Effect on TVA activities, 230
   Hells Canyon case compared with,
   212, 231–232
   Political issue in, 23, 229
   Preferred welfare analysis for, 232–
   233
   Snail role in, 213, 214, 232
Tennessee Valley Authority (TVA),
   214
   Commitment to system of dams,
   230
   Krutilla's work with, 192
   Tellico project and: alternative for,
   212, 216; benefit–cost analysis,
   219–222; environmental impact
   statement for, 213; recreation
   benefits estimates, 222–223,
   224–225; study with Interior
   Department on, 215–216
Tennessee Valley Authority v. Hill, 215,
   233 n. 6
Thompson v. Gallagher, 259 n. 3
Tietenberg, Thomas H., 10
Tinbergen, J., 167
Tobin, J., 175
TVA. *See* Tennessee Valley Authority

Ulph, Alistair, 31
Uncertainty
   Behavior changes with, 16
   Investment and, 11–12
   Natural environment decision
   making and, 61, 63
   Renewable resources
   overexploitation and, 64

Uncertainty (*cont.*)
   Resources allocation and, 5
   Resources valuation under, 15–16
United Kingdom, 33 n. 3, 34 n. 11
User fees, 27, 153, 154
   Locks and Dam No. 26, 250–251
Utility theory
   Attitudes toward equivalent,
      35 n. 16
   Benefits measured under, 63
   Flood control and, 50
   Resources valuation and, 14, 15,
      34 n. 12

van der Tak, Herman, 193
Vaughan, W. J., 186 n. 3
Vietmyer, N. D., 66
Violette, Daniel M., 186 n. 3
Viscusi, W. Kip, 30, 192

Wagner, Aubrey J., 215, 230
Walker, J. D., 145
Walters, Alan A., 17
Water quality
   Ambient standards, 50
   Measuring benefits of, 50
   Research on, 47, 48, 49
   Supply-side concerns, 50
Water Quality Program, RFF, 12, 48
Water Resources Council, U.S., 218,
      222, 230, 246
Water Resources Development Act of
      1986, 202
Water resources projects
   Benefit–cost analysis for, 46, 212
   Management: abuses, 191, agency
      efficiency, 22; political effects, 23
   Measurement of benefits, 49–50
   Research on, 45–46; allocation
      issues, 49, 51–52; effluent
      charges, 49; and environmental
      improvement, 50; groundwater
      contamination, 52–53;

   interdisciplinary approach to,
      53–54; recreation evaluation, 50
River basin development evaluation,
      47–48
Water resources projects, multiple
   purpose
   Costs: problems in concepts relating
      to, 198, 200; proposed sharing
      arrangements for, 201–202
   Efficiency, 6, 22; optimum design
      for, 194; pricing objectives for,
      194–195; prospects for
      achieving, 8–9
   Price system objectives, 119–121
   Ramsey–Baumol–Bradford pricing
      rules, 25–26; 36 nn. 21, 22; 121;
      for cost sharing, 126–128; for
      efficient joint costs allocation,
      129–130; project design
      implications, 128–129; for
      reconciling efficiency and
      revenue objectives, 121–126
   Reimbursement policy for, 22–23;
      cost allocation for, 195; decision
      making reforms for, 205–207;
      effectiveness, 196; evolution,
      195–198; for motivation of
      public programs, 191; nominal
      versus effective rate for, 197–198
Water Supply Act of *1958*, 204
Weisbrod, Burton A., 14
Welfare economics, applied
   Benefit–cost analysis development,
      7, 71
   Evolution, 5
   Externalities, 10
   Focus on efficiency criteria, 7, 8
   Future research for, 28–29, 31–32
   Normative concepts, 71–72
   Relevance, 5–6
   Resource and environmental
      problems' effect on, 27–28,
      33 n. 2

Whalley, John, 127
Whinston, Andrew, 36 n. 27
White, Gilbert F., 30
Wildasin, David E., 127, 129
Willis, William R., Jr., 223
Wilman, Elizabeth A., 18–19, 154,
    160 n. 8, 165, 170, 185
Wilson, Edward O., 59

Wollman, Nathaniel, 45–46

Yaffee, Steven L., 212
Young, H. Peyton, 36 n. 23, 131 n. 5
Young, Robert A., 23, 35 n. 20, 128,
    192

Zeckhauser, Richard, 8, 24, 35 n. 14